Systems Concepts

LECTURES ON CONTEMPORARY APPROACHES TO SYSTEMS

1971 Lecture Series at the California Institute of Technology, Sponsored by the Divisions of Biology, Chemistry and Chemical Engineering, Engineering and Applied Science, and the Humanities and Social Sciences, the Faculty Committee on Relations with Industry, and the Industrial Relations Center

EDITED BY

Ralph F. Miles, Jr.

MEMBER OF THE TECHNICAL STAFF
JET PROPULSION LABORATORY
CALIFORNIA INSTITUTE OF TECHNOLOGY

A WILEY-INTERSCIENCE PUBLICATION

John Wiley & Sons, New York • London • Sydney • Toronto

Published by John Wiley & Sons, Inc.

Library of Congress Cataloging in Publication Data

Main entry under title:

Systems concepts: lectures on contemporary approaches to systems.

(Wiley series on systems engineering and analysis)
"A Wiley-Interscience publication"
"1971 lecture series at the California Institute of Technology."
Bibliography: p.
1. Systems engineering. I. Miles, Ralph F., Jr., 1933– ed. II. California Institute of Technology, Pasadena.

TA168.S865 1973——— 620'.7——— 73-922
ISBN 0-471-60315-5

Printed in the United States of America

10 9 8 7 6 5 4 3 2 1

Contributors

Robert Boguslaw
Professor of Sociology
Department of Sociology
Washington University

C. West Churchman
Professor of Business Administration
School of Business Administration
University of California, Berkeley

Ward Edwards
Professor of Psychology
Head, Engineering Psychology Laboratory
Associate Director, Highway Safety Research Institute
Institute of Science and Technology
The University of Michigan

Ronald A. Howard
Professor of Engineering-Economic Systems
School of Engineering
Professor of Management Science
Graduate School of Business
Stanford University

Robert E. Machol
Professor of Systems
Graduate School of Management
Northwestern University

Ralph F. Miles, Jr.
Member of the Technical Staff
Jet Propulsion Laboratory
California Institute of Technology

Philip M. Morse
Professor Emeritus of Physics
Department of Physics
Massachusetts Institute of Technology

George E. Mueller
President
System Development Corporation

William H. Pickering
Director
Jet Propulsion Laboratory
California Institute of Technology

Simon Ramo
Vice Chairman of the Board
Chairman of the Executive Committee
TRW (Inc.)

Member
Board of Trustees
California Institute of Technology

Henry S. Rowen
Professor of Public Management
Graduate School of Business
Stanford University

SYSTEMS ENGINEERING AND ANALYSIS SERIES

In a society which is producing more people, more materials, more things, and more information than ever before, systems engineering is indispensable in meeting the challenge of complexity. This series of books is an attempt to bring together in a complementary as well as unified fashion the many specialties of the subject, such as modeling and simulation, computing, control, probability and statistics, optimization, reliability, and economics, and to emphasize the interrelationship between them.

The aim is to make the series as comprehensive as possible without dwelling on the myriad details of each specialty and at the same time to provide a broad basic framework on which to build these details. The design of these books will be fundamental in nature to meet the needs of students and engineers and to insure they remain of lasting interest and importance.

Foreword

In 1971 a series of lectures was given on systems engineering at Caltech under the auspices of the Industrial Relations Center, the Divisions of Biology, Engineering and Applied Science, Humanities and Social Science, and Chemistry and Chemical Engineering, and the Faculty Committee on Relations with Industry. Course credit was also given to Caltech students under the course Ae 241, Systems Engineering. A notable array of speakers participated in the series of 10 lectures.

The lectures were very well attended and enthusiastically received. Indeed so great was the response that Caltech and Dr. Ralph Miles, who was responsible for the idea of the lectures and for organizing them, decided that it would be useful to a large number of people to have the material available in more permanent form. This book is the result.

Systems engineering is a necessity for the optimal solution of today's complicated technological problems; this is true even though such problems were being solved many years before the discipline known as systems engineering came into existence. The individuals and organizations that solved these problems were systems engineers, whether or not they knew it. (One is reminded of M. Jourdain, the hero of *Le Bourgeois Gentilhomme,* who discovers in the course of the play that he has been speaking prose all his life.) Yet the creation of a formal discipline, largely as the result of the aerospace efforts during and after World War II, has produced a depth of understanding and skill that has made possible the solution of enormously more complex and difficult technological problems than before.

I am convinced that somewhat the same approach must be applied to the far more complicated social and economic problems that advanced societies face, if we are to make substantial progress in solving them. At the same time, it is clear that systems engineering by itself will not be enough. We do not know enough about the behavior of the components of the socioeconomic problems—the people and the human institutions—to be able to have much confidence in the results of systems engineering approaches to such problems at the present time, no matter how elegant and proven such methods may be in technological areas.

Thus systems engineering is a necessary, but not a sufficient, input for the solution of this class of problems. It is a way of illuminating the facts. The decisions and conclusions in nontechnological areas will have to be reached through the exercise of a great deal of judgment and experience. But, if the facts are not presented and compared and all alternatives exposed through the methods of systems engineering, those judgments, however experienced and able, will have to be made on the basis of faulty data and on the basis of comparisons less precise than they could be.

HAROLD BROWN
President
California Institute of Technology

January 1973

Preface

This book is an edited version of the lecture series, "Systems Concepts for the Private and Public Sectors," delivered in Ramo Auditorium, California Institute of Technology, in the spring of 1971. The 10 lectures of the series were given by renowned experts in the various aspects of systems concepts. The series was attended by members of the Caltech community and by subscribers from the general public. Subscribers to the lecture series represented a diverse spectrum of Southern California technical, business, and government organizations.

There is an abundance of books on systems concepts. The bibliography contained herein lists more than 300. The majority of these books, especially those used as classroom texts, are heavily technique- or analysis-oriented. Little attention is given to such matters as who uses these tools, why they use them, or how successful they have been in implementing their recommendations. Thus the student gains little understanding or appreciation of the difficulties of actually applying systems concepts in realistic environments.

The chapters of this book constitute a set of readings on the application of systems concepts to a wide range of disciplines. The writers discuss the relevance of systems concepts to their professions, what success has been realized, and the possibilities for the future. This book can be considered as supplementary reading to the standard texts for introductory courses on systems engineering or operations research.

Dr. Harold Brown, President of Caltech, provided the initial funding for the "Systems Concepts" lecture series from the President's Venture Fund. The lecture series was sponsored by the Caltech Divisions listed on the title page. Special mention must be made of Robert D. Gray and the personnel of the Caltech Industrial Relations Center for their administrative support in conducting the lecture series.

I am indebted to Professor Francis H. Clauser, Chairman of the Division of Engineering and Applied Science, for the invitation to spend the 1970–71 academic year at Caltech as a Visiting Assistant Professor. I am also indebted to Dr. William H. Pickering and the Jet Propulsion Laboratory for providing my financial support during this period.

RALPH F. MILES, JR.

Pasadena, California
January 1973

Contents

Figures . . . . **xv**

Chapter 1 **Introduction** **1**
Ralph F. Miles, Jr.

Chapter 2 **The Systems Approach** **13**
Simon Ramo

Chapter 3 **The Engineering of Large-Scale Systems** . **33**
Robert E. Machol and Ralph F. Miles, Jr.

Chapter 4 **Decision Analysis in Systems Engineering** . **51**
Ronald A. Howard

Chapter 5 **Divide and Conquer: How to Use Likelihood and Value Judgments in Decision Making** **87**
Ward Edwards

Chapter 6 **Analysis Techniques for Operations Research** **111**
Philip M. Morse

Chapter 7 **Systems Engineering at the Jet Propulsion Laboratory** **125**
William H. Pickering

Chapter 8 **Apollo: Looking Back** **151**
George E. Mueller

Chapter 9 **Planning—Programming—Budgeting Systems** **165**
Henry S. Rowen

Chapter 10 **Systems Concepts in Social Systems** . . **177**
ROBERT BOGUSLAW

Chapter 11 **A Critique of the Systems Approach to Social Organizations** **191**
C. WEST CHURCHMAN

Bibliography **205**

Index **219**

Figures

CHAPTER 1

Figure 1	**Systems Engineering Stages**	**3**
Figure 2	**Hierarchy of Systems, Technologies, and Sciences**	**4**
Figure 3	**The Steps to the Systems Approach**	**9**

CHAPTER 2

Figure 1	**The Telephone System**	**18**
Figure 2	**The Personal Automobile System**	**20**
Figure 3	**A Large American City**	**23**

CHAPTER 3

Figure 1	**A Mathematical Model of a System**	**41**
Figure 2	**Diagrams for the Principal of Superposition**	**42**
Figure 3	**Diagram for a Generalized Linear System**	**43**
Figure 4	**Mathematical Optimization Model**	**44**

CHAPTER 4

Figure 1	**Decision Making (Descriptive)**	**52**
Figure 2	**Decision-Making Using Decision Analysis**	**57**
Figure 3	**Value Assignment**	**61**
Figure 4	**Time Preference**	**62**
Figure 5	**Risk Preference**	**64**
Figure 6	**The Medical Decision**	**66**

Figure 7	**Value of Clairvoyance Computation**	**67**
Figure 8	**A Decision Analysis Model of the Mexican Electrical System**	**70**
Figure 9	**Priors on Material Lifetime**	**79**

CHAPTER 5

Figure 1	**A Single Subject's Probability Estimates**	**95**
Figure 2	**Block Diagram of a PIP**	**97**
Figure 3	**Geometric Mean Odds for Each War**	**100**

CHAPTER 6

Figure 1	**A Queuing Model**	**113**

CHAPTER 7

Figure 1	**The First Jet-Assisted Takeoff (JATO) in the United States, August 6, 1941**	**127**
Figure 2	**The Sergeant, Solid Propellant, Tactical Guided Missile**	**128**
Figure 3	**Four of the Spacecraft Developed by JPL**	**130**
Figure 4	**The Four Systems of a Lunar and Planetary Flight Project**	**133**
Figure 5	**Spacecraft Electronic Parts Count Comparison**	**137**
Figure 6	**Mariner 1969 Project Master Schedule**	**141**
Figure 7	**Cost Trend Analysis for Mariner 1969**	**142**
Figure 8	**Artist's Concept of Morgantown People-Mover Project**	**144**

CHAPTER 8

Figure 1	**Apollo Liftoff**	**152**
Figure 2	**The Apollo Command and Service Modules**	**154**
Figure 3	**The Apollo Lunar Module**	**155**
Figure 4	**Astronaut Working on the Moon**	**157**
Figure 5	**The Five Engines of the First Stage of the Saturn V Launch Vehicle**	**160**

Systems Concepts

LECTURES ON CONTEMPORARY APPROACHES TO SYSTEMS

1

Introduction

RALPH F. MILES, JR.

Member of the Technical Staff
Jet Propulsion Laboratory
California Institute of Technology

Ralph F. Miles, Jr., received his Ph.D. in Physics from the California Institute of Technology in 1963. At the Jet Propulsion Laboratory he was the Spacecraft System Engineer for the Mariner Mars 1969 Project. During the 1969–1970 academic year he was a Visiting Fellow in the Department of Engineering-Economic Systems at Stanford University. During the 1970–1971 academic year he was a Visiting Assistant Professor in Aeronautics and Environmental Engineering Science at the California Institute of Technology. Presently he is the Mission Analysis and Engineering Manager for the Mariner Jupiter/Saturn 1977 Project at the Jet Propulsion Laboratory.

Systems Concepts

Most people have intuitive ideas about the systems approach, or "systems engineering" as it is called in the more technically oriented contexts. Civil engineers have been constructing large systems for a long time—systems such as cities, roads, aqueducts, and pyramids. Today aeronautical, chemical, and electrical engineers design large technically complex systems with complicated man–machine interfaces. Computer programmers, biologists, economists, and sociologists all use systems concepts.

To a large extent these intuitive notions of systems are correct. After all, is systems engineering not just "good engineering," what we have been trying to do all along?

Systems engineering *is* good engineering. And beyond that it is more a change in emphasis than a change in content—more emphasis on defining goals and relating system performance to these goals, more emphasis on decision criteria, on developing alternatives, on modeling systems for analysis, and on controlling implementation and operation.

Collectively, systems concepts constitute a viewpoint and an approach involving the optimization of an overall system as distinct from the piecemeal suboptimization of its elements. In addition, the general class of systems concepts also includes a number of techniques, both methodological and analytical, which are involved in the design and operation of systems.

Webster's unabridged dictionary devotes one full column to the word "system" and its grammatical forms and synonyms. Some of the definitions relevant to our purposes are the following:

- A complex unity formed of many often diverse parts subject to a common plan or serving a common purpose.
- An aggregation or assemblage of objects joined in regular interaction or interdependence; a set of units combined by nature or art to form an integral, organic, or organized whole; an orderly working totality.
- A group of devices or artifical objects forming a network or used for a common purpose.
- An organized or established procedure or method or the set of materials or appliances used to carry it out.
- An organization or network for the collection and distribution of information.

For the purposes of this book, a system is defined as a set of concepts and/or elements used to satisfy a need or requirement. The idea of a system arises when one can associate a need with a capability for satisfying that need. Thus there are many kinds of systems: aerospace systems, sewer systems, administrative systems, cardiovascular systems, systems for gambling, and even systems for beating the system. As shown in Fig. 1, "systems engineering" is defined as the set of concepts and techniques which are necessary to proceed from the original system concept to the creation of the system or, more completely, to the satisfaction of the original need. Thus one can speak of the systems engineering of a planetary mission to Mars or the systems engineering of a more purposeful and efficient judicial system.

There is no clear-cut distinction as to the types of systems for which systems concepts are appropriate. In the spirit of Robert Machol and Ralph Miles in Chapter 3, the class of systems for which systems concepts are relevant have the following properties:

- The system is man-made.
- The system has integrity—all components are contributing to a common purpose, the production of a set of optimum outputs from the given inputs.

- The system is large—in number of different parts, in replication of identical parts, perhaps in functions performed, and certainly in cost.
- The system is complex, which means that a change in one variable will affect many other variables in the system, rarely in a linear fashion.
- The system is semiautomatic with a man–machine interface, which means that machines always perform some of the functions of the system and human beings always perform other functions.
- Some of the system inputs are random, which leads to an inability to predict the exact performance of the system at any instant.

A system need not exhibit all these characteristics in order for systems concepts to apply. Nevertheless, characteristics of this nature determine the degree of relevance and necessity for systems concepts.

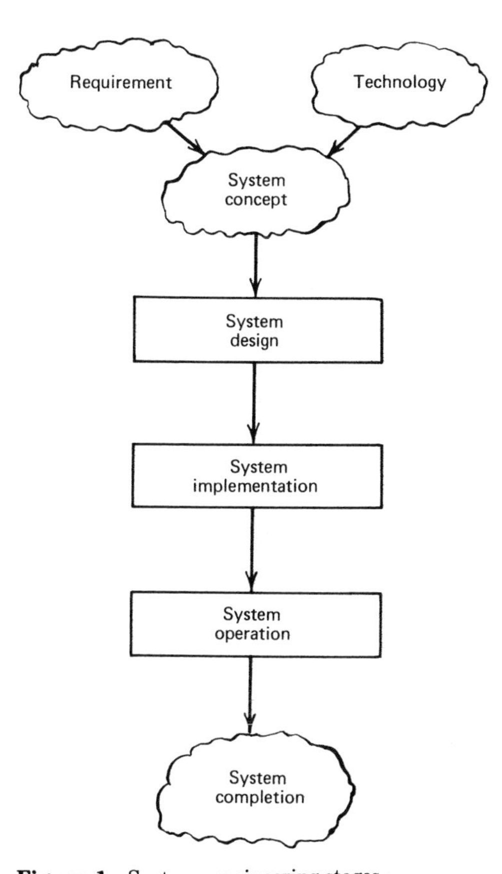

Figure 1 Systems engineering stages.

Figure 2 shows a hierarchy of system types. The base level of the triangle consists of the codification of our knowledge into the basic sciences, wherein we mean to include the soft sciences as well as the hard sciences. The second level of the triangle consists of the technologies, the skills and techniques we have acquired for applying our knowledge. At the third level are the technical systems, which apply, integrate, and manage technologies in a collective effort to achieve a technical goal. Above the technical systems are the civil systems, more directed toward social welfare goals—transportation systems and medical-care delivery systems for example. At the apex of the triangle are the social systems, whose goals are social welfare—management systems, systems of law and justice.

What identifies the level of a particular system? Obvious criteria are related to the amount of technical hardware involved. A more basic and underlying rule says that as one ascends the triangle, the goals or purposes become more and more related to social welfare. There are

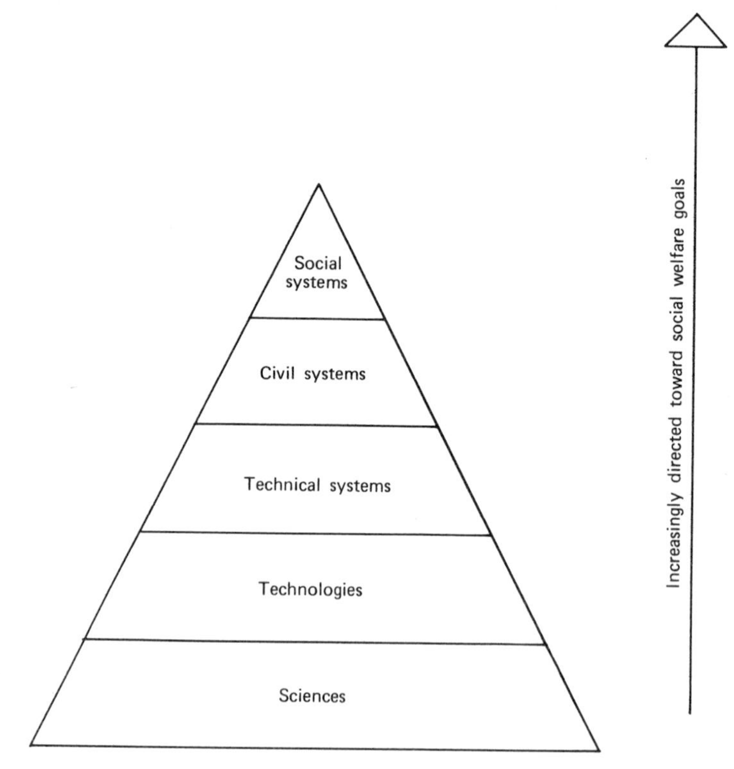

Figure 2 Hierarchy of systems, technologies, and sciences.

no social welfare criteria with respect to the basic sciences. No one asks how an atom ought to be. Social welfare judgment does enter into technology. Should a nation use the scientific knowledge of atom-splitting to build a nuclear power station? An atom bomb?

Purely technical systems are evaluated according to a defined and documented technical specification. The success of a space mission can be evaluated by comparing performance against technical objectives. Civil systems, although strongly pointed toward social needs, can be at least partially evaluated according to technical objectives. Transportation systems can be evaluated partially in terms of technical parameters such as cost-effectiveness and transit times, although the design of a civil system is fraught with social consideration. Whom shall the system serve? Who shall bear the costs? What levels of safety shall be required?

Social systems are primarily designed to enhance social welfare—governments instead of anarchy, law and order instead of lawlessness and chaos, education to remove ignorance. Easily measured social indicators often do not give a true picture of the effectiveness of a social system. Student-teacher ratios and cost of facilities do not in themselves measure the knowledge and skills imparted by an educational system.

Underlying every technology is at least one basic science, although the technology may be well developed long before the science emerges (e.g., glassmaking). Overlying every technical or civil system is a social system which provides purpose, goals, and decision criteria. Ultimately, of course, all systems involving people are embedded in a social system.

Solid-state physics is one of the basic sciences underlying the technology of transistor making. A machine to automatically make 10,000 transistors a day, or a factory to produce 10,000 transistor radios a day are examples of technical systems. A network of radio stations in an underdeveloped country, broadcasting educational material to illiterate peasants, would be a civil system. Whether an underdeveloped nation should allocate limited resources for such a system would be a question to be posed within the framework of the social system.

Overview of the Book

The organization of the book roughly breaks down into a set of chapters on theory and a set of chapters on application. There is not a clean break, however, primarily because all the authors have been involved in the practical application of systems concepts.

Chapter 2, by Simon Ramo, "The Systems Approach," provides an overview of systems concepts and their wide range of applicability. Systems engineering became accepted as a necessary activity in the design of complex systems around the time of World War II and has since been used in the design of many modern systems in industry, transportation, communications, and government. In Chapter 3, "The Engineering of Large-Scale Systems," Robert Machol and Ralph Miles discuss some of the guiding principles for systems engineering.

Designing and implementing systems requires that a large number of decisions be made—decisions concerning complex problems in complex environments. Decision analysis attempts to structure and analyze these complex decision problems in such a way so as to maintain the same logical relationships that would exist in an elementary problem. There is a long-standing question as to whether complicated real-world decision problems are amenable to modeling and analysis. It is undeniably true that the universe within which real-world decisions are made is richer in complexity than any model could ever hope to capture. Nevertheless, decisions *are* made and resources *are* allocated. The evidence is that all this occurs not on the basis of the complex universe but on the basis of relatively simple models of decision situations. What decision analysis purports to do is to incorporate into the analysis of real-world problems logical preciseness and the correct expression of preferences between alternatives. In Chapter 4, "Decision Analysis in Systems Engineering," Ronald Howard presents the decision analysis cycle and describes the roles of probability, utility, value, information, time, risk, and uncertainty in decision analysis.

Chapter 5, by Ward Edwards, "Divide and Conquer: How to Use Likelihood and Value Judgments in Decision Making," complements the preceding chapter on decision analysis in considering psychological concepts implicit in decision making. In any contrived, hypothetical decision situation, it can be shown that decisions arrived at through formal logic are superior to decisions made on the basis of intuition. Edwards develops the thesis that these same techniques of formal logic can be applied *in toto* to real-world decision situations and that this approach appears once again to be superior, though no proof is possible.

Large systems are inextricably complex. Existing techniques of analysis have been reshaped and new techniques developed to meet the needs for systems analysis. In Chapter 6, "Analysis Techniques for Operations Research," Philip Morse discusses some of the analysis techniques and mathematical models useful to operations research and shows how they have been applied in a practical manner to systems problems.

The aerospace industry has made spectacular use of systems engineering—world-wide communication systems employing satellite links, Apollo systems that fly to the moon and return, spacecraft that travel hundreds of millions of miles to distant planets. The modern procedures for managing, planning, implementing, and operating complex aerospace systems were first developed on communication systems and ballistic missile systems. The procedures were extended to all military systems and were later incorporated by NASA.

The Jet Propulsion Laboratory has made a significant contribution to the development of systems engineering through its management of, and participation in, lunar and planetary projects. In recent years the Jet Propulsion Laboratory has applied its systems engineering expertise to civil systems problems such as medical-care delivery systems, environmental control, crime prevention, and the design of transportation systems. In Chapter 7, "Systems Engineering at the Jet Propulsion Laboratory," William Pickering presents the systems concepts that have been used in the design of lunar and planetary missions and discusses how these systems concepts have been transferred to the design of civil systems.

In Chapter 8, "Apollo: Looking Back," George Mueller discusses various problems encountered in the design and implementation of the Apollo program. He makes the point that in spite of the enormity of the Apollo program—the utilization of vast resources, the number of highly skilled people involved, the size and complexity of the systems—ultimately the resolution of the major problems rested with a very small number of people making extremely difficult decisions in the face of great uncertainties.

Planning-programming-budgeting as a management system originated in the Department of Defense during the 1960s. The essential aspects of this management system are: a careful specification and a systematic analysis of objectives; a search for the relevant alternatives, the different ways of achieving the objectives; an estimate of the total costs of each alternative; an estimate of the effectiveness of each alternative, of how it comes to satisfy the objectives; and a comparison and analysis of the alternatives. In Chapter 9, "Planning-Programming-Budgeting Systems," Henry Rowen discusses the success to date in applying this system in government.

In Chapter 10, "Systems Concepts in Social Systems," Robert Boguslaw presents the idea that social systems are designed by the components of the system. Man is the basic component of a social system, and the system exists to enhance his welfare. Requirements and constraints flow up the system hierarchy as well as down. Thus the social system designer is as much a negotiator and an arbitrator as he is a designer.

In the final chapter, "A Critique of the Systems Approach to Social Organizations," C. West Churchman reviews the history of organizational theory and the attempts to view these organizations as social systems to which systems concepts should be applicable. He concludes by stating a need for the application of the systems approach to systems analysis itself and by predicting that the systems approach of the future will incorporate what he describes as a dialectical learning process.

The Systems Approach

Quoting from Simon Ramo in Chapter 2:

> "The systems approach is a technique for the application of a scientific approach to complex problems. It concentrates on the analysis and design of the *whole,* as distinct from the components or the parts. It insists upon looking at a problem *in its entirety,* taking into account all the facets and all the variables, and relating the social to the technological aspects.

In applying the systems approach, the systems-oriented person recognizes that needs or problems originating at one level invariably have contributing factors at higher levels. Thus one should attempt to view the immediate needs or the exacerbating problem within a larger context. What are the factors that created the need or caused the problem to arise? Is the problem complete in itself or is it merely a manifestation of a larger, more fundamental concern? More police, supported by more exotic technology, could reduce the crime rate in city slums but would do little to get at the underlying economic and sociological problems. To minimize the occurrence of cancer, a medical doctor tells his patients not to smoke cigarettes. A behavioral scientist, in taking a systems approach, attempts to alleviate the psychological and sociological factors that led the patients to smoke.

To apply this approach to a systems problem of any consequence requires a vast wealth of knowledge and the interaction of a diverse number of talents. Thus the use of the much heralded "systems team" comprised of specialists from all the relevant technologies.

Using the systems approach and doing systems engineering involves solving a lot of problems, and for this reason it is valuable to examine these systems concepts within a problem solving context. John Dewey stated the essence of problem solving some sixty years ago when he asked:

1. What is the problem?
2. What are the alternatives?
3. Which alternative is best?[1,2]

Every human being, be he the president of a multibillion dollar corporation or an aborigine of Western Australia, goes through life solving problems. Indeed, it can be said that life presents itself as a sequence of problems, terminating with one you can't solve! So there is absolutely nothing new about problem solving for human beings. What the systems approach purports to do is to logically structure the problem-solving methodology.

Dewey's formulation has today, largely through the influence of the communications and aerospace industries, evolved into the celebrated "systems approach." As shown in Fig. 3, a problem, need, requirement, or goal is quantified in terms of objectives that the system must satisfy and criteria that can be used to rank alternative systems. A process of system synthesis takes place in which a set of alternative systems are generated. Each of these systems is analyzed and evaluated in terms of the stated objectives and design criteria. The "best" or "optimum" system is then selected and implemented. Of course, in practice the process is extremely iterative, with results from later stages fed back to early stages to modify objectives, criteria, system options, and the like.

How would you use the systems approach? Let us assume that you have a situation which is concerned with a need or involves a problem. The systems approach asks you to do the following:

The Systems Approach

1. Goal definition or problem statement
2. Objectives and criteria development
3. Systems synthesis
4. Systems analysis
5. Systems selection
6. Systems implementation

Figure 3 The steps to the systems approach.

[1] John Dewey, *How We Think,* D. C. Heath, 1910.

[2] R. A. Johnson, F. E. Kast, and J. E. Rosenzweig, *The Theory and Management of Systems,* McGraw-Hill, 1967, p. 280.

1. Express your understanding of the situation in a logical, coherent manner: some words, a picture, mathematics if that is possible.
2. Develop a set of objectives and criteria that the system must satisfy in order to achieve the goal. If the situation involves a problem, state the characteristics that will exist when the problem has gone away.
3. Develop alternatives to resolve the situation; not just one, but a set of alternatives from which you can pick and choose.
4. Examine and analyze each of your alternatives with respect to your goals and criteria. Select the alternative which you prefer, and implement the solution.

The success of the systems approach up to this time indicates that the process works well when the system objectives can be clearly formulated, and when the required technologies and sciences are sufficiently mature. The objectives for the Apollo project can be clearly stated: "To place a man on the moon and return him safely before the end of the decade." During the 1960s four major system technologies required for the Apollo implementation reached maturity: large launch vehicles, vehicles for operation in space, a system for trajectory analysis and orbit determination, and a system for communications.

Two major factors inhibit the successful application of this design process to civil and social systems. The first and most fundamental is that system objectives and system criteria can rarely be clearly stated. How does one choose between two transportation systems: one which gives a fast, bumpy ride and one which gives a slow, smooth ride? By what criteria does one discern the optimality of an educational system?

The second inhibiting factor is the lack of maturity in the required technologies and sciences—the soft sciences. The "descriptive" man yet remains to be described. In addition to unknowns in physiological man, knowledge of psychological man is in a rudimentary state. The lack of this fundamental information results in one library, optimally designed for use, which has no windows and a second library, also purported to be optimally designed, which has many windows. Even less understood is the subject of what man ought to be—"normative man." Yet this question is implicit in the design criteria of every social system.

Systems engineering draws on all the concepts of the basic sciences and disciplines. The present state of the art for systems engineering is that there now exists a well-demonstrated methodology for integrating technical disciplines into technical systems. Civil and social systems now lie on the frontiers of systems engineering. It is possible that civil and social system designers may experience only limited success until the psychological and social nature of man is better understood.

It thus appears that the systems approach works well when certain conditions are satisfied. These conditions are met for the design of technical systems, but only partially so for the design of civil and social systems.

In some ways it is unfortunate, though almost inevitable, that the modern concept of the systems approach has evolved out of highly sophisticated technical programs. It has come into being burdened with the technical jargon of the aerospace business and cloaked with the mystique of computers and mathematics. Yet the true nature of the systems approach has the purity of simplicity and the believability of common sense.

The systems approach is just plain common sense in that each concept, each step, is the reasonable thing to do. The value of the systems approach is that it allows you to bring all these common-sense ideas together in concert to focus on the resolution of complex problems in complex environments.

The systems approach will not solve problems for you. Only you can do that. What the systems approach will do is permit you to undertake the resolution—your resolution—of a problem in a logical, rational manner. You are the one who must ascertain that a problem or a need exists. You are the one who must develop alternatives. You are the one who must develop the criteria for selecting a suitable alternative. The systems approach will not do any of these things for you.

Psychologists say that we human beings yearn for uniqueness, for the right to be an individual, a very special individual. The systems approach will give you this opportunity in a very rational framework. The systems approach will allow you to express your individualism rationally when you identify your problem, your alternatives, and your decision criteria.

2

The Systems Approach

Wherein a cure for chaos presents itself.

SIMON RAMO

Vice Chairman of the Board
Chairman of the Executive Committee
TRW Inc.
Member
Board of Trustees
California Institute of Technology

As chief scientist for the ICBM program, Simon Ramo coordinated the scientific efforts of one of the nation's largest technical programs. An organizer of Hughes Aircraft Company's electronics and missile operations and, subsequently, cofounder of Ramo and Wooldridge, which later merged into TRW Inc., of which he is Vice Chairman of the Board and Chairman of the Executive Committee, Ramo has founded corporate organizations that are in the forefront of computers, missile systems, electronics, transportation, and city planning. He is the author of several texts on science and technology, and has recently written two books on the application of science to social problems, *Cure for Chaos* and *Century of Mismatch.* Ramo is a Trustee of the California Institute of Technology, and is a founding member of the National Academy of Engineering.

The Need for a Systems Technology

There are two major areas of our society where advanced technology has been applied with noticeable success—business and national defense. In the free enterprise sector, private capital readily plants the seeds and grows the fruits, and the resulting products represent a rough correlation between the capabilities of industry and the requirements of society. The competitive market and the profit incentive match the demand for a product with the resources necessary to get it developed and produced. The system is far from perfect, and it does not cover everything. But it works well enough, and a substantial fraction of the potential of technology is exploited for society's needs.

In the case of national defense, when the products required do not flow automatically out of the free enterprise system, technological resources are mobilized under government sponsorship and control. The results are costly and the activities are usually controversial, but the government-industry-science relationship does get the job done.

In contrast, there remains a large area of national need where technology has hardly been brought to bear: control of natural resources, transportation within and between cities, air and water pollution abatement, city development and redevelopment, education, public health. These problems, which we shall call "civil systems," tend to have certain characteristics in common. They involve unarticulated goals. They are all large and complex. Solutions appear to be extremely expensive, and typically they require the use of sophisticated technology. No improvement is possible without the cooperation of many semi-autonomous groups which are not accustomed to cooperating. The problems generally cannot be solved by the development of a single product or service. New concepts, new apparatus, difficult adjustments and novel functions for people, untried interconnection among them—all are needed.

Even to analyze these problems and come up with practical and meaningful proposals is not easy. This alone usually requires some social progress. Civil systems problems therefore do not lend themselves to quick or complete solution by free enterprise. For the private sector to furnish answers, there would have to be some kind of market, some identifiable group of consumers. But if you want to unpollute a river into which several cities pour refuse and industrial waste, who will finance the development of the equipment—who is the customer? A market relationship between industry and society at large does not exist in this sector, as it does in the military area. Consequently, technological solutions elude us.

The subject of this chapter is the application of science to this field of civil systems. Within the next decade, there is reason to believe a cure for chaos will be developed.

For one thing, we are seeing now an awakening of public interest in these problems. Thinking people are beginning to ask questions. If we can give men good air to breathe on the moon, then why not in our cities? If technology permits us to fly to Europe in a couple of hours, why can't we apply technology to get us to the airport in less than that time? If our knowledge of energy is such that we can destroy society in a matter of minutes, why can't we use this energy to desalt the water in the ocean? If we can record and analyze the voltage of a little gadget in a guided missile a thousand miles away, why can we not provide really superb medical instrumentation for bed patients in our hospitals?

At the same time, concerned citizens realize that solutions will be difficult, slow, and expensive. We do not have the time or the money to embark on grossly wrong solutions, then disembark and start over. We need all the logic and creativity and objectivity we can possibly muster.

Meanwhile, side by side with the growth of public interest, a new methodology has been maturing. This methodology, called the "systems approach," has been an essential ingredient in the successful application of science to military and space systems. It is beginning now to be recognized that the systems approach is also well suited to attacking the civil systems type of problem.

The systems approach is a technique for the application of a scientific approach to complex problems. It concentrates on the analysis and design of the *whole,* as distinct from the components or the parts. It insists upon looking at a problem *in its entirety,* taking into account all the facets and all the variables, and relating the social to the technological aspects.

When solutions are envisaged, they are expressed in the form of a detailed *system,* combining men and machines, assigning functions to each, specifying the use of materiel and the pattern of information flow, so that the whole system represents an optimum ensemble for achieving a particular set of goals.

The systems approach applies logic, wisdom, and imagination on a sophisticated technological level. It is often quantitative and always objective. It makes possible the consideration of vast amounts of data and of numerous, often conflicting, considerations. It spells out the interactions among the elements of complex real-life problems, recognizing the need for careful compromises, for "tradeoffs" among competing factors such as time versus cost. It uses simulation and mathematical modeling, when applicable, to predict performance *before* the entire system is brought into being. By insisting on an examination of the total problem—the goals, the criteria, the costs, the benefits, and the penalties—it seeks to disclose what we can expect to get and what it will cost us, and it makes feasible the selection of the best from among many alternatives.

It would be an overstatement to call the systems approach "new" intellectual discipline. In some ways it is a familiar old concept. Note the word "systems" in telephone systems, weapons systems, the Federal Reserve System, the subway system. The old meaning is still the new meaning; the word "systems" connotes the whole, the combination of many parts, a complex grouping of men and machines.

The notion of attacking a large problem in an organized way is certainly even older. When the Sphinx was built, and the Roman roads and the London Bridge and the Panama Canal, there had to be in every instance someone whose job was to relate the current technology to the objectives, to the social environment, and to the available resources. We did not arrive at our telephone and electric power distribution systems by the random, accidental dropping from the sky of pieces of apparatus that just happened to work well when connected up together.

You don't even have to be a professional to use the systems approach. When any of us has a problem—preparing a budget, choosing where to live or what job to seek, designing a chair or a house, selecting a route for a trip—in every instance it is well to be logical, to use the wisdom we possess, to consider objectively all the factors involved, and to recognize that there are many alternatives.

What makes this approach appear new today? Partly the accelerated development of the tools of systems engineering in recent years. This has resulted, in great part, from the need for this kind of methodology in the highly complex and costly defense and space programs. Large electronic computers make possible the information processing that is basic to good quantitative analyses of the pertinent parts of real-life problems. Next to the skilled human brain, the computer is the most important tool of systems analysis.

For another thing, there are now for the first time a substantial number of professionals who are well-seasoned in interdisciplinary problems. They know how to relate one facet of technology to another and also to all the nontechnological factors that characterize practical problems. We have begun to develop good "systems teams," combines of individuals who have specialized variously in mathematics, physics, chemistry, the many branches of engineering, sociology, economics, political science, physiology, business finance, education, and government. The systems team knows how to pool its knowledge and attack a problem from many directions. This is worth stressing, because it is quite often assumed that the systems approach refers to a group of narrowly specialized engineers, skilled in the details of technology, but ignorant of behavioral science, sociology, and other nontechnological considerations. Such a concept of the systems approach is completely erroneous. Individual members of a systems team may be specialists, but the team as a team must have breadth that matches the problem or, by definition, it is not a *systems* team.

Good Systems and Bad

Within a few years we shall substantially exceed a one-trillion-dollar gross national product. This means that during the decade of the 70's the total price of all the products and services we buy will be ten trillion dollars, give or take a few billion. Fully 10 percent of the total, one trillion dollars, will represent efforts in civil systems—transportation, urban development, pollution control, medical and educational facilities, and so forth. The true value to society of this enormous expenditure can be altered greatly depending on whether or not the effort is properly chosen, well organized, and efficiently operated. The application of scientific method and sound technology, with prime attention to the interface of technology and society—which is what the systems approach seeks to effect—will exert tremendous leverage. The cost of applying the systems approach will be small compared to what it will accomplish.

To understand this "leverage" effect, let us take some familiar systems and observe the contrast between the "piece-meal," disorganized approach and the systems approach, and measure the results of each.

The Telephone System

The telephone system, for example, is not just the instrument we hold in our hands. It is a closely integrated network of men and equipment stretching for thousands of miles, arranged so as to make possible the immediate interconnection of any two points in the whole system. It includes installation, repair and maintenance services, phone books, bills, cables, satellites, emergency services, central stations, pay phones, and conference calls. It works so well today, because from the very beginning it was seen as a system and not merely a congeries of unrelated, independent elements (Fig. 1).

Let us imagine that instead the telephone matured in a kind of ad lib, helter-skelter fashion, through a series of cumulative oversights or accidental wrong decisions that overlooked the system requirements. Suppose some enterprising craftsman starts making telephones in his little backyard workshop and sells one to you. You hire an electrician to connect your instrument to that of a friend some distance away with whom you would occasionally like to speak, or to your physician's office, or to a business firm. Just as the government provided and maintained roads, so the city will put up some poles along the sidewalks and leave it to your electrician to string his wires between

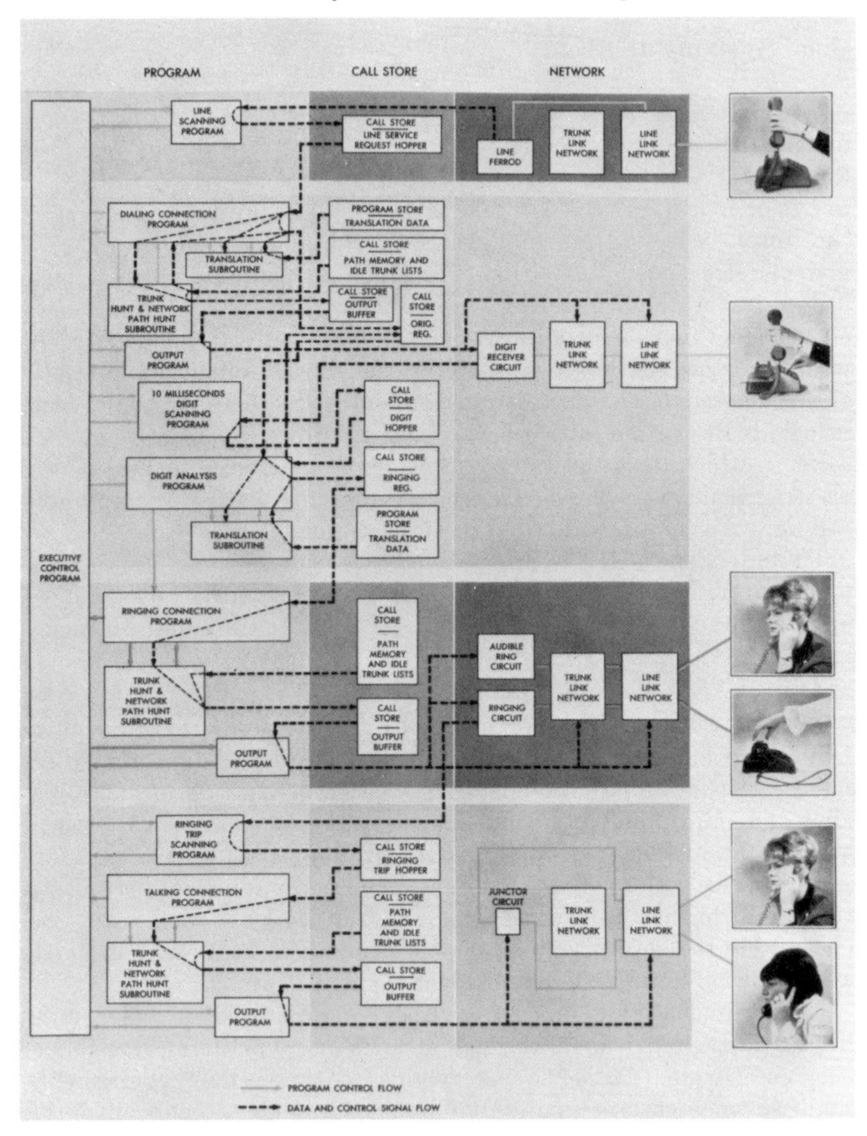

Figure 1 The telephone system. The telephone system works well today, because from the very beginning it was seen as a system and not merely a congeries of unrelated, independent elements. (Copyright 1965, Bell Telephone Laboratories, Incorporated. Reprinted by permission, Editor, Bell Laboratories RECORD.)

them. When there are too many wires on the pole and the pole begins to crack and some wires fall down—well, that's no different from the roads that are already congested and full of potholes. Like your car, sometimes your phone calls will get through and sometimes not.

As you develop the habit of talking to your friends and business associates, you and they keep stringing up more wires. Of course pretty soon someone would notice that the number of lines this takes limits telephone service. For instance, if you had a thousand people who wanted to be able to talk, each one to every other one, then you would need hundreds of lines going in and out of all the buildings—nine hundred ninety-nine from each one, if you can picture it.

Eventually, an enterprising businessman might think to set up a "central station" and let everyone connect to him for a fee, so that he could somehow take care of connecting you to the others when you want to talk. Sooner or later such central operations would spring up all over the country, with several competing stations in the larger cities. With this kind of system, you might still be able occasionally to arrange a long distance call to a few highly restricted places, just as you can get across Manhattan by car occasionally on a busy day, and maybe even find a parking place. But you wouldn't be able to hear very well and it would be terribly expensive.

Sooner or later the practical requirements of our technological society for fast, flexible, economical, reliable communication would inevitably have forced the telephone service into good systems practices. But the point is that the systems approach was seen here to be necessary from the beginning.

The Personal Automobile System

By contrast, consider the personal automobile "system." The system, of course, comprises not only cars but roads, spare parts, gasoline, red lights, driver licensing, casualty insurance, parking lots, traffic courts, and many other subsystems without which the over-all system could not operate. But it was not designed as a harmonious, integrated system. It just grew.

And as a result it doesn't work. Vehicles capable of traveling a hundred miles an hour are normally found crawling, and often frozen into motionless congestion. The system wastes millions of man hours each day in traffic, saps the nervous energy of the drivers, and pollutes the air we breathe. One million people have been killed by automobiles in the United States since World War II. If the personal automobile system had been designed from the outset as a rational ensemble of people and equipment set up to answer the requirement for personal transportation, we would have saved time, money, and lives (Fig. 2).

It is to be emphasized that although each of us has a personal choice of which automobile to purchase, this does not mean we can choose the over-all system of personal transportation. The automobiles available are no doubt properly chosen by their manufacturers (otherwise they would not be successful producers), but they are chosen to fit into the system that exists.

The random, casual, ad lib approach that accompanied the personal automobile into American life was fine for many years. Some roads existed, the garages and gasoline and spare parts could come along gradually, with everything, generally speaking, led by supply and demand. By now, however, the over-all system is a complicated mismatch of uncoordinated, independent aspects which overlap in such a way that to straighten out the system is almost impossible. In large

Figure 2 The personal automobile system. If the personal automobile system had been designed from the outset as a rational ensemble of people and equipment set up to answer the requirement for personal transportation, we would have saved time, money, and lives. (Los Angeles County Air Pollution Control District.)

cities, where the problem is especially conspicuous, the personal automobile cannot be considered independently from mass rapid transit, or air pollution, or the many other aspects of urban development and redevelopment.

In such a case, unless we attack problems as systems problems, unless we surround the total situation and recognize that all the factors must be respected, we have chaos, not just lowered performance. Let one musician play whatever comes into his mind, and at worst you may have something you do not care to listen to. But let each of one hundred excellent musicians pick his own notes in complete disregard of the others, and the result will be noise. It is not just a desire for efficiency, time saving, or economy that causes us to want harmony in our civil systems. It comes down to a choice between something and nothing at all, between music and cacophony, between a semblance of order and utter, total, terminal chaos.

The Air Transportation System

If you are a frequent air traveler, you cannot help being struck with how close we sometimes seem to come to a breakdown in civil aviation. A slight increase in the magnitude of the basic variables—number of passengers, number of planes trying to land, amount of baggage to be handled, amount of paper work at the ticket counter, deterioration of weather—and the badly belabored system is threatened with paralysis. An air transport system, we must emphasize, is much more than the airplane. It is getting to the airport, baggage, blind landing radar, air traffic controllers, communications systems, maintenance crews, ticket agents, food preparation, training schools, and regulatory laws. At a certain level of activity, the interaction of these factors goes beyond the threshold of compatibility, and breakdown occurs.

Let us take an example. A plane arrives thirty minutes late for one reason or another. In a complex situation that is close to the threshold, a compounding effect sets in. That flight has missed its scheduled location in the pattern. It is now difficult to take care of it—in the extreme case, almost impossible. Other flights are automatically delayed. The next take-off of that same plane has to be postponed, not by thirty minutes but by two hours, while it waits for the maintenance crew that was busy preparing other planes. Passengers discovering a delay of their flight, shift to competitive airlines. The reservation apparatus is jammed, interfering with still other flights. The baggage people are busy trying to reroute everybody's bags.

It is like a dog suddenly invading a busy freeway. The first driver puts his brakes on suddenly. This requires the driver behind him, who is following closely, to stop even faster to keep from running into driver number one. Driver number three behind is caught with an even greater challenge, and drivers four, five, six, and seven simply do not have a chance. They have to pile into each other.

A question naturally arises. Could this kind of run-away congestion that brings a system to a halt have been averted by safety margins in the design of the system? Just as the cars on the freeway should not follow so closely, perhaps we should not schedule air operations so tightly. But it is difficult to arrange that cars do not follow closely, if we have to handle great peaks of traffic within reasonable costs; we cannot snap our fingers and have lots of parallel lanes when the traffic gets high. It is even harder to build leeway into an air system. There are so many overlapping variables that even with what might look like luxurious surplus capacity in each element, the system can still be very close to saturation in actual operation, unless a careful and creative job is done of anticipating and working out all the interrelationships. You can be generous in most aspects and still have no reserve. And if you do succeed in building excess capacity to take care of the sharp peaks, you still must expect deep valleys, when the facilities will be used only lightly, and when much expensive manpower will be idle. More importantly, you must severely limit the total amount of air traffic.

The air transportation system in the United States has reached the point where we cannot afford to maintain it without better systems design. Building luxuriously high over-capacity is precluded not only by cost but by the way our cities happen to be designed and how our population happens to live. From now on, the air transportation problem must be looked at as an entirety. All the elements must be accorded weight and attention, individually and together, so that a consistent, logical, optimized system will evolve, and resources will be used to the fullest. Systems design will analyze the frequency of different kinds of situations, and show clear, quantitative tradeoffs between cost and effectiveness. Innovative procedures will be evaluated, in the interest of getting the largest possible reserve by optimizing all the relationships.

Good systems design in the air transport system is not easy to arrange. There are many autonomous groups involved. But unless integrated design dominates in the future, chaotic conditions may become commonplace.

The American City

Time is running out in other sectors of our society as well. If you want an example of imminent disintegration resulting from inadequate handling of a systems problem, then you have it in the typical large American city. Housing, mass transportation, race relations, education, waste removal, air pollution, water supply, electric power reliability, distribution of food and materiel, crime, medical care, rats—the list is overwhelming. It is increasingly clear that we are approaching a threshold of intolerability in many communities where the whole system will fall apart. The city that first reaches that threshold will commence to be abandoned. Either that, or it will cease to be an organized community.

A city *is* a system, whether or not we choose to regard it in that light. If we choose not to, then it will simply be a bad system. (Fig. 3).

Figure 3 A large American city. A city *is* a system, whether or not we choose to regard it in that light. If we choose not to, then it will simply be a bad system. (Los Angeles Department of Water and Power.)

Now, in the next few decades, as we add a hundred million to the population of the United States, we shall undoubtedly be building new cities as well as extending and modifying existing ones. Some are going to be designed, or redesigned, by the systems approach—with proper and competent consideration of all the factors and their relationships. These designs will not be perfect, but they will be superior, socially and economically. They will offer better living, higher income, and greater social stability. The flow of people and things, and the access to vital services, will mesh with the work and life patterns of the people who live there. Such communities will offer economic advantages for businesses and industries situated within them, compared to the decaying cities. In the latter, the situation will become increasingly difficult to handle.

The Hospital System

The systems approach is becoming vital today for still another reason. Without a good systems analysis as a first step, in many cases it is not possible to identify the *components* of the solution so they can be made available. These components, which systems design brings together to meet the problem, are of three kinds: (1) *equipment* and materiel; (2) *people,* trained in specific jobs with spelled-out functions and procedures; and (3) the right kind of *information* stored and flowing, so that the people and things know what to do and where to be to make the whole system operate. If the systems analysis is not done first, then the components are never identified. So no one builds the machines, or trains the people, or designs the information flow.

Let us consider an example of this. In a large hospital, test data on patients must be provided to the physician in a timely and reliable fashion. Now it may just be that the best way to do this is by an electronic system with the acquisition, storage, and display of test information. Such a system would probably include electronic consoles to be interrogated by the physician at convenient locations, perhaps even in each patient's room. Similar devices are widely used today by stockbrokers and airline reservations personnel, so we know, technologically, that such devices can readily be designed for hospital use as well.

But precisely what *kind* of box is right for the hospital service? What must be its storage capacity? Its display capacity? Where should it be located and how should it be connected up? The answers cannot come solely from a designer of electronic boxes working with no knowledge of hospitals. Nor does it make sense for some entrepreneur to drop in on a hospital and ask a few stray physicians or administrators what they would like to have and how much they are willing to pay for it. No one knows enough to say something authoritative about this specific single piece of an as yet undesigned medical system.

Before you can design such a component, you have to design the system. You cannot build something on vague or guessed-at specifications and then sell your product into a non-existent market. It would be like writing the oboe part for the second movement of a symphony not yet composed.

The Northeast Corridor

Take another example of a project where the systems approach has to lead before we can get the components. The federal government has now tackled the problem of moving people about in the most highly populated section of the country, the Northeast Corridor between Boston and Washington, where a very large fraction of the total population and industry of the United States can be found. Unless a superior transportation system is developed for this region, frequent major interruptions, severe bottlenecks, social disruption, and a net drag on the nation's economic growth will result.

There is not a prayer of even denting the problem without a thorough systems analysis. Until the whole program has been worked out in sufficient depth so that the interrelationships of all the factors become clear, we will never get the new trains, engines, tracks, computer controls, rights-of-way, feeder roads, ticket booths, and bond issues that are required.

As things stand now, the manufacturers of apparatus do not have a clear market for the sale of their products. They won't even really know what to build until there has been a good and accepted system design. Such a design must consider not only the patterns of how people live and work and transport themselves to do both, but how these are going to change in the future, and, indeed, how they will be changed by the choices made of future transportation systems. Of course, one cannot consider railroads or buses without reference to airplanes and personal automobiles. The cost and effectiveness of various combinations of ground and air, public and private transportation must be estimated, along with the effect of the total environment.

What good is it, for example, to be able to move rapidly between two adjacent cities by monorail if the system hasn't been designed with enough completeness to allow for getting from your home to the station in a sensible way. If a personal automobile must be used to do that, then there had better be a way to get the car parked when you get to the station.

It will also be necessary to consider routing high-speed ground transportation through certain rural areas not so heavily populated today. This will enhance the population buildup in those areas, and minimize the more costly modifications that might have to be made in existing cities if instead the population built up there. But the rural route may not be as good for those in existing cities. What is the tradeoff here?

Even setting up criteria to compare one solution with another is a tough job. A transportation system serves the people who live in the area, but the people will live there to some extent because of the existence of the transportation system. Without investigating the interactions of some of these basic parameters, how can one judge sensibly how big passenger vehicles should be and at what speed they should travel? Or determine what the over-all capacity of the system should be, or how to enhance any aspect of performance? Or specify the physical description of all the devices, controls, communications, terminals, ticketing systems—the whole works?

Fortunately, the leadership of the Boston-New York-Washington complex has understood from the beginning the usefulness of the systems approach. While carrying on systems analyses and designs, they have brought potential component suppliers into the fraternity of intelligence that is considering the whole system.

The Computer on the Beat

Another reason for the growing importance of the systems approach to many social problems is that the solutions of these problems inevitably include a significant ingredient of advanced technology. At this time of the history of man, almost any large and complex problem in any field of endeavor is generally best solved by arranging men and machines in an intelligent configuration. This is so whether our task is designing a highway, a stock exchange information system, an insurance company's data bank, a police department's communication and control system, a medical center, a university library, or a whole city.

In recent years, a particularly pertinent area of technology, namely electronics, has greatly expanded its utility in this respect. Through electronic means we are able to remember and handle so much more information all at once, still keeping everything straight, that systems to accomplish tasks can be envisaged, built, and operated that would have been completely out of the question a decade or two ago.

Jet travel, for example, would be quite impossible if airlines could not provide virtually instantaneous computerized reservations. If everything had to be written out by hand and moved by direct human-to-human communication, it would take longer to book a seat than to fly across the country.

The Northeast Corridor transportation system is another example. A sensible system analysis here requires evaluating all the possible ways of moving people hundreds of miles along the ground and in the air, including not only conventional means but also monorails, automated tunnels through the ground, huge new land cruisers on special railed highways, and various other devices using advanced technology. Such a system cannot be analyzed soundly without going into the new technology to come.

Similarly, any modern city trying to create a sensible traffic system to get the largest number of people through the streets in a minimum of time with the greatest of safety now must consider all the latest technology, including the possibility of synchronizing lights by a computer connected with traffic sensing devices.

Even a modern police department today can use electronics to improve its procedures. Deploying available officers in available cars so as to get the greatest crime prevention coverage per man on the force is a crucial function of police headquarters. Monitoring the location of police cars is an easy task for a computer. Careful systems engineers, working closely with experts on police operations, can program the computer so that when some cars move in response to emergencies, the remaining cars will be immediately directed to move to new strategic locations. These locations will be chosen on a statistical basis to provide the best new spread. If this were done well, it would give the effect of more officers and automobiles on patrol at the same cost. Whether this kind of system would actually improve police operations may be arguable, but good systems engineering is required if the police want to be in a position even to judge the possible value of new technology.

The best example, however, is one we have used before for other purposes; namely, the medical center. Computers and other electronic devices can revolutionize many aspects of hospital design and operation. Results from an electrocardiograph or a metabolism measuring instrument can go directly into a computer system for storage and, later, instant retrieval. Simultaneously, the fact that the test was made can be registered in the accounting department, and a bill can be issued from there automatically. A patient's heartbeat, temperature, breathing, and other conditions can be detected by monitoring devices in his room and the results displayed electronically in a central command post where continuous observation can be arranged over a number of rooms. Even the general appearance of the patient can be monitored on a continuous basis by television devices. The computer can do a better job of scheduling test facilities; the line in front of the X-ray machine can be shorter. Electronic inventory control systems for drugs and other supplies can ensure the most reliable availability of all needed supplies with the least investment in inventory.

Applying the Systems Approach

In planning any system, it is necessary to call in people who are experienced in applying the most advanced technology. Competent systems teams must include such technologists, just as they must include economists and psychologists, in order to evolve a good systems design. But it is a mistake to assume that all good technologists are always capable of applying the systems approach well. There are plenty of nontechnological problems in every system. Moreover, it takes flexibility, imagination, and experience to mesh the technical with the nontechnical. It is only when this is done well that the systems approach is properly applied.

Insofar as systems analysis merely involves the use of objectivity, logic, and wisdom, it is inexcusable not to use this approach to every problem. But the formal, professional use of the systems approach—the engaging of a team of expert systems engineers, the dignifying of both the problem and the methodology that is implied when the systems approach is consciously brought into play—can be misunderstood and oversold, leading to unsatisfactory solutions for important problems. There is art as well as science in professional systems work.

Take the question of how big a problem one really should surround with the systems approach. You start off to plan a new hospital, for instance, and you have to consider changes in the practice of medicine, changes in governmental policies (medicare and the like), changes in the affluence of the society, the population growth in the area, and the general health of the population. There are so many factors that seem to affect the system directly that the first thing you know, if you really want to be complete, you find yourself trying to predict the total life pattern for the entire population of the area. But the area is tied to surrounding areas and to the nation, and the nation is tied to the world. Such an approach, carried to an extreme in seeking to cover all aspects of the problem, is absurdly ambitious and impractical.

Surrounding a problem *too* broadly, trying too hard to be absolutely complete, is very poor systems engineering. After all, optimizing is one of the key concepts in the systems approach, and it is hardly an optimum attack on a problem to embrace so huge a definition of it as to require a billion dollars and twenty years just to assemble all the facts.

A skillful systems team closes in on the problem and concentrates on those factors which are closer in and most significant. As the issues are categorized, those which are seen to be less direct are dealt with to a more and more superficial degree, partly for lack of adequate knowledge about them and partly because otherwise you will make no headway at all.

But are there not some cases where the systems approach is indicated, but where the problem is inherently too big to permit the systems approach? For example, the entire economic system of the United States.

Though the problem is big, it is indeed a system—a constellation of people, things, information, and money all involved together in an extremely complex, interconnected network which determines the economic life of the nation. The system is there, designed or not, analyzed or not.

Now it has apparently become accepted practice for the federal government to tinker with this system. We all want freedom from severe business cycles, from recession and inflation, from unemployment. The government can influence these things in many ways. It can modify government expenditures, control interest rates and the money supply, alter taxes, put controls on the production of many things, stockpile materiel, modify tariffs, assign rates to utilities, subsidize industrial and service operations, determine the marketability of gold, allocate resources, set minimum wage rates, locate government facilities in chosen parts of the country, set routes and fares for airlines, influence labor management negotiations, to name a few.

So the economy is a system, and it is being directed, to some extent, in an effort to meet various objectives. But whenever something goes wrong in the system, then immediately there are wide differences of opinion among the experts as to what steps the government might take and what the effects of these steps would be.

Now if we had a very accurate mathematical model of the entire system on a computer, we would be able to work out ahead of time exactly what to do, and we would know its effect before we did it. We could then choose the best course, because everyone would accept the logical, optimized answer that results from this full application of the systems approach.

Alas, this is not possible. First of all, we are unable to assemble even a small fraction of the pertinent facts. This is, after all, a pretty big system. We would almost have to get down to the individual contributor, each man and machine. But even if we could conjure up by magic all the required statistical data, our knowledge of the system principles is inadequate. We do not fully understand the basic economic relationships between inflation, unemployment, prices, wages, interest rates, taxes, and the other key factors.

On the other hand, while we do not have all, we have part of the needed information. We do understand *some* of the relationships among the main factors. Economists have made considerable progress in the last hundred years. Now the computer has arrived on the scene, and it has been applied to the handling of important economic data. Attempts are being made, with considerable success, to relate at least some of the main factors to one another, to set up mathematical models, and to use these models to predict, for example, next year's Gross National Product as a function of varying assumptions of governmental and other actions. These attempts are being taken more and more seriously by leaders in economic theory and by those who need the results in their work.

Skillful systems teams, therefore, can now provide us with solutions to *parts* of the problem with a degree of logic and objectivity that we should not fail to make use of merely because we cannot yet control the entire economic system. We have a system that is vital to us, and that we are already forced to influence by action. We should certainly prefer logic to illogic, facts to guesswork, objectivity to emotional or political hunches, and begin to act accordingly.

Fortunately, most of the real-life problems that need solutions today are considerably smaller than the economy of the United States. As a practical matter, it is possible to isolate pieces of most big problems, and arrive at conclusions that make the systems approach pay off. If we are looking at interurban transportation in the Northeast Corridor, we regret we cannot predict completely the life habits of people in the year 2000. But it is useful to compare five or six ways of moving people a hundred miles in that area, and see whether, in any set of circumstances that we can imagine, one of these approaches will be superior to the others. Then we can at least use that understanding as a guide toward making better decisions.

From this point of view, there is no such thing as a system that is too big for the systems approach, just as there is no epidemic that is too big for the useful practice of medicine. Though total success is as yet beyond us, we should not throw away the tools that can give us partial aid.

The systems approach is a bottleneck breaker. The more it is used, the easier it is for it to get used. It is often a first step in determining how much money something will cost. It helps to articulate goals that might have been only vaguely understood before. Because it is logical and quantitative, it provides comparisons and tells you what you will get for what you pay. It discloses how a proposed system would fit into the existing operations of society.

Most civil systems problems are complex and controversial, and involve conflicting objectives and conflicting interests. Many times nothing can be done until a new plateau of objectivity is somehow attained, and solid specifics of what the choices are and what their consequences would be are made clear. This is just what the systems approach is all about. The systems approach therefore encourages action in a society where many people must work together before action is possible.

In addition, systems analysis promotes organizational innovation in our social structure, which is usually required before we can cope with our unsolved problems. Any good systems solution includes procedures for coordination and control. These in turn suggest the type of organizational structure which will most effectively implement the solution. Successful systems engineering therefore leads toward the creation of some new organization that no one could otherwise have designed.

Thus, for example, a really good systems analysis might show how a major river can be unpolluted. Certain actions will be indicated, at specified costs, with certain benefits and corresponding drawbacks. If a majority of the affected population accepts this program, it will be obvious that they must create some sort of organization to implement it. But the very details of the systems analysis will show what this organization must be like, what powers and funds it will need, how it will have to work with some existing authorities and supersede others, and so on. Unless the systems analysis has been accomplished, it is not going to be clear to anyone what new legislation, bond issues, etc., are needed. The systems approach may be the bottleneck breaker here in pointing the way to action.

The systems approach is still a new, little understood, and relatively underused discipline. It may well be another ten years before it is applied on a large enough scale to alter the balance between technological advance and lagging social maturity. But in a decade or so, the public, the Congress, local governments, and leaders in industry and science may all be convinced of the value and importance of the systems approach. At about that time, we will face a new bottleneck: a shortage of good systems engineers, including, of course, the non-technologist members of the systems team—the economists, political scientists, psychologists, and sociologists. The work is difficult. Assembling technical and non-technical specialists into working groups with the wisdom and imagination required cannot go forward as rapidly as is desirable.

Still, it is pleasant to imagine a time when the only thing that retards the use of logic, objectivity, and all the tools of science is a lack of enough trained professionals. That will be the beginning of a golden age. Once most people are wedded to a logical and objective approach to social problems, the world will be a lot better, and science and technology can be used to the fullest on behalf of society.

BIBLIOGRAPHY

Ramo, Simon, *Cure for Chaos,* David McKay Co., Inc., 1969.

Ramo, Simon, *Century of Mismatch,* David McKay Co., Inc., 1970.

Ramo, Simon, "New Dimensions of Systems Engineering," *Science and Technology in the World of the Future,* A. B. Bronwell (Ed.), Wiley, 1970, pp. 127–146.

3

The Engineering of Large-Scale Systems

ROBERT E. MACHOL

Professor of Systems
Graduate School of Management
Northwestern University

RALPH F. MILES, JR.

Member of the Technical Staff
Jet Propulsion Laboratory
California Institute of Technology

Robert E. Machol is a Professor of Systems in the Graduate School of Management, Northwestern University. He is the editor of *System Engineering Handbook* and *Information and Decision Processes* and is a coauthor of *System Engineering,* an introduction to the design of large-scale systems. He was formerly the President of the Operations Research Society of America, the Head of the Department of Systems Engineering at the University of Illinois, and was the Vice President of Conductron Corporation. For Ralph Miles' biography, see p. 1.

Systems Definition

Many books on systems engineering have been written, and in each the authors have been faced with the problems of defining the term "system." There is no unanimity in these definitions, and it does not appear that any universally acceptable definition is likely to emerge in the near future. The most that can be asked is that the author delineate the area about which he is discoursing so that the reader can have an indication of their common areas of interest.

To pick one example, Johnson, Kast, and Rosenzweig define a system as follows: ". . . an array of components designed to accomplish a particular objective according to plan."[1] This very general defini-

[1] R. A. Johnson, F. E. Kast, and J. E. Rosenzweig, *The Theory and Management of Systems,* McGraw-Hill 1967, p.113.

This chapter is based on material in "Methodology of System Engineering" by R. E. Machol in *System Engineering Handbook,* McGraw-Hill, 1965, pp. 1-3 to 1-13. Portions are reprinted with permission of the publisher.

tion is, in fact, too general for our purposes. It encompasses entities which we will not regard as systems or for which our systems concepts have little relevance.

The following characteristics are typical of those that restrict the class of systems to those of interest.

1. The system is man-made, and it incorporates equipment, computer software, procedures, and the like. This eliminates anthills, river basins, universes, and many other interesting "systems."
2. The system has integrity—all components contribute to a common purpose, the production of a set of optimum outputs from the given inputs. What this purpose is, how we define optimum, and even the nature of the inputs will often be unknown at the start of the system design process, and their elucidation will be an important part of the task. Rigorously applied, this criterion would exclude cities and, indeed, nearly all social systems.
3. The system is large—in number of different parts, in replication of identical parts, perhaps in functions performed, and certainly in cost; such things as the ignition system in an automobile are thus excluded.
4. The system is complex, which is here taken to mean that a change in one variable will affect many other variables in the system, rarely in a linear fashion; in other words, the mathematical model of the system will be complicated. This eliminates systems which are merely large, such as a bridge or highway (apart from considerations of traffic flow).
5. The system is semiautomatic, which means that machines and computers perform some of the functions of the system, and human beings perform other functions of the system. The large, completely manual system (pyramid building with slaves or a large data bank maintained by clerks) is eliminated because it is too inefficient for our interests. The large, completely automatic system is eliminated because it either does not exist, or else exists only as a subsystem of a larger system with an essential man-machine interface.
6. The system inputs are probabilistic, which leads to an inability to predict the exact load or performance at any instant. In some cases the rate of input is predictable (e.g., in an automatic factory), but even here there are difficulties in design because of unpredictable variability in such things as environment and raw materials.
7. Many systems, especially the most complicated, all-encompassing systems, are competitive. In military systems a rational agent (the enemy) is trying to destroy or reduce the effectiveness of the

system; in business systems ordinary competition, or in public service systems cheating or mere noncooperation, have similar effects.

It must be understood that no one of these characteristics is necessary and no subset is sufficient. A system, much like beauty, lies in the eye of the beholder. There is no generally accepted definition which will separate systems from nonsystems. A decade ago, the title of "system," within our context, was usually construed to apply only to technical systems, but today the term has been extended to civil and social systems.

An example of a system which would appear to be amenable to straightforward analysis might be a transportation system for intercontinental travel. One could imagine many alternative systems, ranging from ships to rockets, but let us assume that criteria such as minimum travel time, cost effectiveness, and technical feasibility dictate that the optimum alternative uses a large jet airplane of the Boeing 747 class.

Now, the system does not consist of just the 747 airplane; pilots are needed to fly the plane, stewardesses are needed to help the passengers, a ground crew must prepare the plane and load the baggage. In addition, the plane must take off and land on airfields. While in flight the plane occupies an airlane, hopefully to the exclusion of other planes. All this requires facilities, personnel, and procedures. Passengers do not appear as if by magic, in one-to-one correspondence with the number of seats on the plane. Thus there must be a reservation system, a subsystem of our transportation system, to insure that an adequate, but not too many, number of passengers will be present for the flight.

All these activities must be managed by an organization that operates and maintains the plane, and ensures that personnel, supplies, and facilities are available in the right number at the right time. The proper management of this operation requires that estimates of future requirements be made years in advance, based on a mass of data concerning past operations and on indicators of future trends.

Finally, there must be organizations, desireably few in number, concerned with the overall resources invested and the overall return obtained. At one level this involves the stockholders and the management of the airline. At a higher level it involves the national government, or a consortium of governments. It is they who must provide the overall objectives and decision criteria for the system and make the difficult trade-offs between public and private investment, convenience versus efficiency of operation, user versus nonuser considerations, and profit versus safety and reliability.

Thus what started off to be a straightforward technical system with seemingly well defined interfaces on closer inspection becomes a subsystem treaded through many larger systems, with system objectives so complex that we are left with limited prospects of realizing complete answers to even such basic question as, "Do the benefits equal the resources expended?"

Systems Design

Consider the design of a system. Someone hands you $100 million and asks you to make a system to control air traffic, or to connect 70 million telephones by direct dialing, or to perform some other function. What do you do next?

The designer of such a system immediately confronts a dilemma, because the problem of designing a large-scale system is overwhelming, if it is attacked all at once; yet, if the attack is piecemeal, it is unlikely to be successful. The hope of the designer is that the problem can be subdivided in such a way that the parts can be handled somewhat separately, and that ultimately these parts can be rejoined in a straightforward manner to form the total system. This subdivision can take place simultaneously in a number of conceptually different ways. In particular, there are:

1. The logical steps of systems design.
2. The chronological phases of systems design.
3. The functions of the system.
4. The components of the system.

In Chapter 1, Ralph Miles lists the steps of the systems design process as: (1) goal definition or problem statement, (2) objectives and criteria development, (3) alternative synthesis, (4) systems analysis, (5) systems selection, and (6) systems implementation. These are the logical steps of systems design, but rarely can they be performed in this order. Logically, one must formulate the problem before one solves it. In fact, one performs both functions simultaneously throughout the systems design process. Because the problem cannot be adequately formulated until it is well understood and cannot be well understood until it has been more or less solved, the two are inseparable. Thus the design of any system is extremely iterative, with the designer proceeding from problem formulation to solution to problem reformulation and so on, with each cycle producing a more refined, better understood, and, in principle, more optimal system.

The chronological phases of systems design can be ordered as definition, design, implementation, and operation. The definition phase involves an analysis of the requirements and selection criteria, a generation at the systems level of a range of feasible alternatives, and an analysis of the best alternatives. The conclusion of the definition phase comes with the selection of one systems alternative, and with a gross understanding of the implications of the selected alternative with respect to performance, cost, schedule, risk, required technology development, system lifetime, interfaces with other systems, and so forth.

The design phase starts with the product of the definition phase, a grossly defined system, and proceeds to define, design, and analyze the system down to the level such that all documentation exists for the complete creation of the system. The implementation phase brings the system into being. This phase includes the procurement of parts and materials, fabrication and assembly of hardware, coding and validation of computer software, and training of personnel. The implementation phase ends with the system level tests or review processes which are required to certify the system for operation. The operations phase starts with the first application of the system to its stated purpose and continues through to the final phase-out of the system at the end of the life-cycle.

While these phases logically follow one another, in practice, there will always be some overlap. The degree of overlap that is permitted in the chronological phases of a system life-cycle is derived from trade-off considerations between speed of system implementation and a desire for cost-effectiveness and risk-minimization. Large military-systems procurements have been carried out under both philosophies, and there are advocates and adversaries of both sides of the issue.

It is often convenient to subdivide a system along functional lines. NASA divides its deep space communication network into a set of six functional systems which cut across international, administrative, and facility boundaries. The six systems are: tracking, telemetry, command, simulation, monitor, and operations control. Hospitals divide their operations into functionally differing services such as surgery, internal medicine, pediatrics, and so on.

Perhaps the most tangible way of subdividing a system is by its respective components. Nevertheless, even here care must be exercised to achieve a breakdown which will aid and not detract from the designer's efforts. The natural and desired boundaries should lie at points that minimize the interaction across the interfaces. It would be sheer folly to subdivide the design efforts for a complicated electronics system by part type, for example, a resistor subsystem, a capacitor subsystem, and so on, even though such a subdivision might make the most sense for the parts procurement phase.

These four subdivisions—logical, chronological, functional, and component—often merge in the system design process, as demonstrated in the following sentence: "Determine the performance requirements (logical subdivision) for the design (chronological subdivision) of a transmitter (component subdivision) for the communications link (functional subdivision)."

Given that one does proceed in this manner—subdividing the problem in convenient and productive ways, then going through a process of synthesis and analysis, then recombining to form a new system which is more optimal than the results of the preceeding iteration—the question then is raised, "How does one stop?" Or does this iteration go on ad infinitum, with each cycle yielding one more increment of optimality? What one should realize, of course, is that the system under design must be viewed within the context of a larger system—a system which includes the resources being expended in the design process. Now the answer to the question becomes clear. The iterations of the design cycle cease when the marginal return of an iteration no longer exceeds the marginal cost of the iteration. Thus we conclude this section with the seemingly contradictory statement that no optimally designed system is optimal!

Principles of Systems Design

The fundamental principle of system design is simply to maximize the expected value. Obviously this requires considerable interpretation in any particular case, but at least the expected value has a succinct and well-understood definition. Where one has the choice of supplying too much of something (resulting in excessive cost) or too little (with the possibility of a penalty if it proves inadequate), this rule gives a guideline, and this kind of thing is done continually in the design process.

Thus we have "trade-off analyses," in which, for example, for an airplane we might compare increased takeoff power (for a potential increase in payload) versus decreased fuel economy (for a potential decrease in payload). In such a two-parameter analysis it is conceptually simple to find a maximum; in a complicated systems situation we would have to trade also with dozens of other parameters, with pairwise comparisons being totally inadequate. This leads to "cost/effectiveness studies," in which we attempt to maximize the effectiveness of the system (or its expected value) for fixed cost, or to minimize the cost for fixed effectiveness. Because it is generally impossible to find a single number which realistically represents the effectiveness of a complex system, there is a good deal of subjectiveness, as represented by judgment, as well as objectiveness, as represented by analysis, in systems engineering.

The principle of suboptimization states that optimization of each subsystem independently will not lead in general to a system optimum, and that improvement of a particular subsystem actually may worsen the overall system. Since every system is merely a subsystem of some larger system, this principal presents a difficult, if not insoluble, problem,—one that is always present in any major systems design. We will discuss this point further in one of the following sections.

The principle of centralization refers to centralization of authority and decision making, that is, to centralization of information as distinguished from material. Most organizations are built on the principle that routine inputs are handled at a low echelon, with higher echelons being informed so they may veto specific decisions or change policy on general decisions, if it is appropriate (this is sometimes called "management by exception"). Nonroutine decisions are passed to higher echelons for decision. This decision hopefully establishes a policy so similar decisions in the future will become routine. A difficulty arises only when the speed required for making the decision exceeds the speed with which the information may be communicated to the higher echelon and the decision made there and transmitted back. Thus, as speeds of communication and decision-making increase, the disadvantages of centralization decrease. With the improvements in computers, communications, displays, and theories of decision-making, the optimum in the continuum between centralization and decentralization moves more in the direction of centralization in our complex systems.

The principle of events of low probability is related to the fact that no system can be all things to all people, all of the time. The principle states that the fundamental mission of a system should not be jeopardized, nor its fundamental objectives significantly compromised, in order to accommodate events of extremely low probability. Yet one frequently hears: "the most trivial detail may be the key to the entire intelligence picture; therefore, the system must be able to store and process every conceivable intelligence input" in spite of the fact that the resulting system is too complex to be workable. In other words "the soldier in the foxhole may have urgent requirements for the airborne reconnaissance information; therefore, the entire data processing system must be airborne, and provision supplied for air drop of finished data to the front lines" in spite of the fact that the resulting allocation of weight to airborne data processing equipment will seriously compromise the reconnaissance performance. The system engineer can sympathize with the soldier in the foxhole and the commander who is sensitive to his needs, but he should insist on reasonable compliance with the fundamental principle of maximizing the expected value of the system.

In many systems, a compromise is possible: the system can be designed to handle most events automatically, and to sound an alarm which calls for manual intervention when an uncommon event occurs which is beyond its capablities. For example, an automatic mail-sorting system would throw out, for manual sorting, those letters which were not of standard size, shape, or location of address. Such a system might handle 95 percent of the mail automatically, at a cost much less than that of 100 percent manual handling and enormously less than 100 percent automatic handling. Similarly, when you reach a wrong number through (automatic) direct-distance dialing, you simply call an operator for (manual) rectification of the error.

Models for Systems Engineering

A model, in principle, is a substitute for the real thing. Models are used as tools to gain knowledge through analysis and as a means of conveying information. A model may be used in lieu of the real thing for any of a number of reasons: economy—it may cost less to derive knowledge from the model, availability—the model may represent a system which does not yet exist or cannot be manipulated, information—the model may be a convenient way to collect or transmit information. Models form an important part of systems concepts because economy, availability, and information are all important factors in the design and analysis of large, complex, and dynamic systems.

There are many ways of classifying models which are useful for systems design and analysis. Three that we shall consider are simulation/symbolic, structural/empirical, and descriptive/normative.

Simulation models replicate a system in function or form. A drawing can be said to be a simulation of a system because it looks like the system. An analog computer simulates a system in that its parameters have the same time history as the system. A system test program is a simulation of the operations phase in which the functional and environmental interfaces of the system are simulated, and the system is tested in its operational modes.

Symbolic models have no physical or functional resemblance to the system. Symbolic models use ideas, concepts, and abstract symbols to represent a system, as expressed in the form of words, graphs, or mathematics. The documentation for a system would represent a symbolic model, as would the mathematical representation of the system operation.

Models may be both simulation and symbolic, depending on the point of view. Computer programs of complicated systems, which are obviously symbolic models, are often called simulation models because they simulate many parameters of the system and, in some cases, may actually duplicate the software portions of the system.

Another classification of systems models describes them as structural or empirical—structural, if the parameters and functional relationships of the model have direct correspondences with the elements of the system; and empirical, if the parameters of the model are adjusted to give the model correspondence to the system, but the parameters in themselves bear no relationship to the elements of the system. A Taylor series expansion of system response or a linear regression analysis of data would be examples of empirical models.

An important classification is the differentiation between descriptive and normative models. A descriptive model describes a system without making any assessment of the system's value or of the system's performance. Normative models describe a system as it would be if it satisfied some criterion of optimality. The design requirements or contract specifications for a system capability would represent a normative model of the system. The degree to which the descriptive model of a system corresponds to the normative model would be a measure of the optimality of the described system. Normative models provide goals for systems design and systems operation, whether or not, in fact, they are achieveable.

A simple mathematical model of a system can be represented by a transformation between an input set and an output set: $y = S(u)$ (Fig. 1).

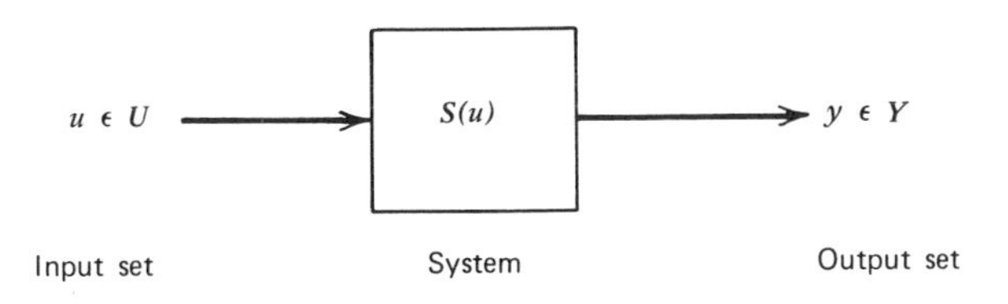

Figure 1 A mathematical model of a system.

This simple model will suffice to demonstrate an important concept in mathematical modeling, the division of math models into linear and nonlinear systems.

A system model is linear if and only if the input and output sets of the system can be represented as elements of linear vector spaces and the principle of superposition holds for the system relationships between the input and output sets, that is, the output of the system for a set of combined inputs is equal to the sum of the outputs for the individual inputs. More precisely, for a linear system with two inputs u_1 and u_2:

$$y = \alpha y_1 + \beta y_2$$

where

$$y_1 = S(u_1)$$

$$y_2 = S(u_2)$$

$$y = S(u) = S(\alpha u_1 + \beta u_2)$$

and α, β are scalars. In graphical form, this is shown in Fig. 2.

The mathematical techniques for linear models form an important part of systems analysis for a number of reasons. Many systems are linear, and many nonlinear systems exhibit a linear response over a restricted range of the system parameters. Linear systems theory is most general in application, while nonlinear systems analysis must often be applied on an *ad hoc* basis, with results which cannot be generalized to a broad class of systems. Finally, linear systems theory is conceptually easier to understand, and in fact forms a starting point for many nonlinear techniques.

If the internal nature of a system varies as a function of time or of prior inputs, then the nature of the system is called the *state* of the system. The state of an electrical circuit can be characterized by spec-

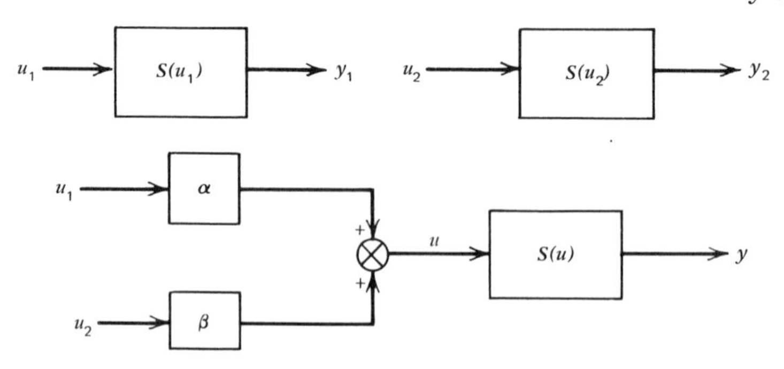

Figure 2 Diagrams for the principle of superposition.

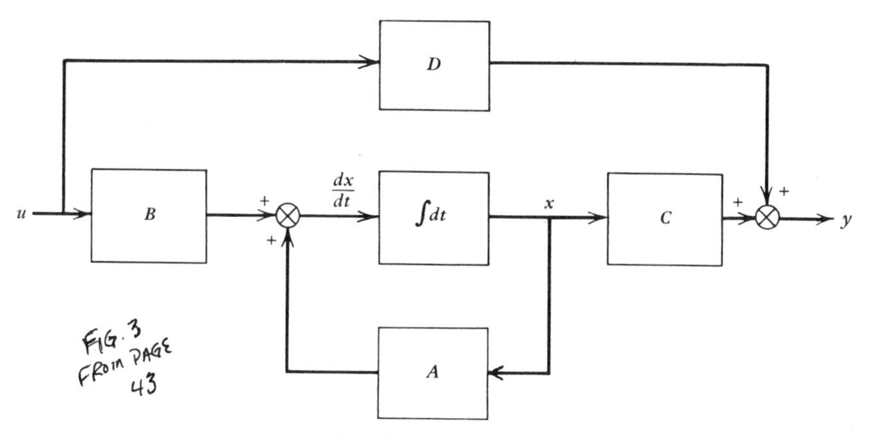

Figure 3 Diagram for a generalized linear system.

ifying the amount of energy in the circuit, that is, charge on capacitors and current through inductors. The state of an economy might be characterized by the available workforce and functioning facilities and businesses. The output of such an economy might be characterized by gross national product.

A general model of systems with internal states can be obtained by expressing the time rate-of-change of the internal state as a function of the internal state and the applied input:

$$\frac{dx}{dt} = f(x, u)$$

$$y = g(x, u)$$

where x is the system state, u is the input to the system, and y is the output or the observable quantity of the system. For linear systems, these general system-state equations reduce to

$$\frac{dx}{dt} = Ax + Bu$$

$$y = Cx + Du$$

These linear equations look much more imposing when they are drawn in graphical form (Fig. 3). Equations of this form can be used to analyze many different systems with a spectrum of differing inputs.

Such mathematical equations have wide applicability, ranging from aerospace control systems and chemical process plants to models of economic and social systems.

The *raison d'etre* of systems design is optimization, and a class of models useful for this purpose are those referred to as "mathematical programming" or "mathematical optimization" techniques. All of these techniques incorporate some model of the system, along the lines we have just discussed, plus a mathematical statement of the criteria for optimality. This statement is called a value function or an objective function. The optimization process may attempt to maximize value, a benefit-cost ratio, output, reliability; or to minimize cost, risk, inputs; or some combination of these—whatever the system designer perceives to be the measure of the "best" system (Fig. 4).

The simplest, though inelegant, optimization technique is to examine all of the possible system alternatives, one-by-one, in an "exhaustive" search, and then to select that system or set of system parameters which produces the optimum value for the objective function.

Where defined, logical relationships exist between the system alternatives, all the analysis tools of mathematics may be brought into play—calculus, variational methods, Lagrange multipliers, and so forth. Elegant computer programs have been developed to search efficiently through large volumes of parameter space for optimum solutions.

Several new optimization techniques have been specifically developed to deal with the optimization problems of large systems. Linear programming models represent systems by a set of linear inequalities of the form:

$$\sum_j a_{ij}x_j \leqq b_j$$

and the objective function by an equation which is linear in the "decision" variables, x_j:

$$Z = \sum_j c_j x_j$$

The optimization process consists of selecting the x_j's to maximize or minimize Z subject to the constraints of the linear inequalities. The "simplex" method for the solution of linear programming problems was developed in 1947 by George B. Dantzig and his associates for the U.S. Air Force on Project SCOOP (Scientific Computation of Op-

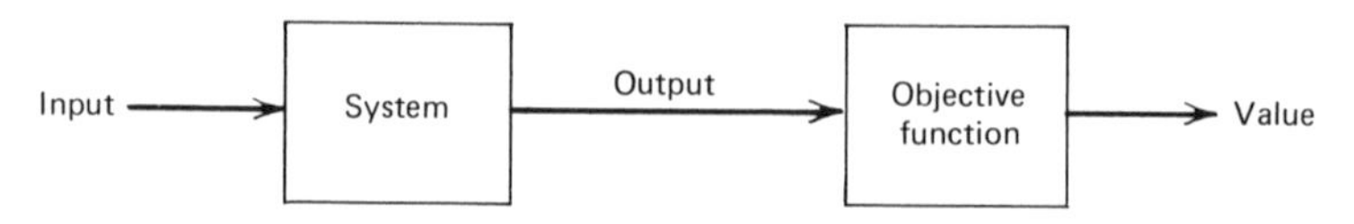

Figure 4 Mathematical optimization model.

timum Programs). Linear programming techniques were originally applied by the Air Force to such diverse areas as contract bidding; balanced aircraft, crew training, and wing deployment schedules; scheduling of maintenance overhaul cycles; personnel assignment; and airlift routing problems.

Another technique, dynamic programming, was originally developed by Richard Bellman in the early 1950s at RAND. Dynamic programming models are a tool for analyzing multistage decision processes. One can imagine systems situations where a series of decisions must be made concerning a number of activities, or a time-sequenced set of decisions must be made concerning a single activity. Dynamic programming models make the following basic assumptions:

1. The returns from different activities can be measured in a common unit.
2. The return from any activity is independent of the allocations to the other activities.
3. The total return can be obtained as the sum of the individual returns.

These assumptions can be expressed as

$$\text{optimize:}\quad Z = f_1(x_1) + f_2(x_2) + \cdots + f_N(x_N)$$

subject to the constraint

$$x_1 + x_2 + \cdots + x_N = x_0$$

where the x_j's are the resources to be allocated and x_0 is the total resource available.

Often, systems problems can be modeled as a flow of something through the system. This flow may be automobiles through a traffic pattern, goods in a manufacturing plant, oil through pipelines, activities through a scheduling chart, or the like. These problems are analyzed with the aid of network flow models, where the network consists of interconnecting nodes and paths. One important technique is called PERT (Program Evaluation and Review Technique). PERT was developed by the U.S. Navy as a technique for preparing project schedules, and for assessing the progress of the project with respect to the schedule. PERT was first used on the Navy's Fleet Ballistic Missile Project—the Polaris project. The management of the Polaris project was complicated by the fact that there were many tiers of contractors and subcontractors. It was extremely difficult to understand the impact of problems in one area on other areas and on the overall project schedule. PERT made an important contribution to the resolution of these scheduling problems on the Polaris project.

There is a fundamental difference in the use of models in science and engineering. In science, a model is the final product. A model in science represents the distillation of all the knowledge about a certain phenomenon. In engineering, where the purpose is not codification of knowledge but the achievement of an objective, models are only a means to an end. How much modeling effort should a systems engineer undertake, and how detailed should the systems model be? This is a resource allocation problem. The systems engineer should continue to develop the systems model until the marginal benefit of improving the model falls below the marginal cost.

The Systems Viewpoint

Systems engineering is more than a knowledge and application of principles of systems design and systems modeling concepts. In what follows an attempt is made to give the reader a feeling for the point of view which makes systems engineering different from classical engineering.

The heart of the matter lies in the complexity of the system and in the danger of being unable to see the forest for the trees. The designer must somehow deal with the various subsystems and component parts in such a way as to optimize the cost/effectiveness of the overall system—which means avoiding the dangers of suboptimization. The word "suboptimize" was coined in 1952 by C. J. Hitch, and the following example is taken in part from his article.[2]

An excellent study, one of the classics of operations research, was performed during World War II on the optimum size of a merchant ship convoy.[3] The problem was the sinking of United States and Allied merchant ships by "packs" of German submarines in the North Atlantic Ocean. There is, of course, never enough data for such problems, because of the statistical variability in such things as sightings and sinkings and the numerous questions of luck and skill involved. However, the researchers were able to show (with what most system engineers would agree was reasonable confidence) that the number of merchant ships sunk when a convoy was attacked by a given pack of submarines was independent of the number of merchant ships in the convoy, but inversely proportional to the number of escort vessels

[2] C. Hitch, "Sub-optimization in Operations Problems," *J. Op. Res. Soc. Am.*, vol. 1 (1953), 87–99.

[3] P. M. Morse, and G. E. Kimball, *Methods of Operations Research*, Wiley 1951.

(such as destroyers) in the convoy. Furthermore, the number of submarines sunk in such an encounter was proportional to the number of escort vessels. It follows that the payoff, chosen as the ratio of submarines sunk to merchant ships sunk, varies as the square of the size of the convoy (assuming that the same ratio of escorts to merchant ships is retained). The recommendations from this study were put into effect, and the number of merchant-ship sinkings decreased drastically, contributing importantly to the winning of the Battle of the Atlantic, and consequently to the winning of the war. In fact, the decrease was even more dramatic than predicted; the submarines were so ineffective in the North Atlantic that they were transferred to more profitable missions elsewhere.

This celebrated problem has been the subject of a number of postwar studies, and it now appears that the change in tactics (increasing convoy sizes) was probably right, but for many of the wrong reasons. In fact, the study as described above is remarkable for the number of errors which have been made from the systems viewpoint.

In the first place, if one really believed the above conclusions, he would recommend taking every bottom available to the Allies and putting them into a single giant convoy. This is clearly ridiculous (in systems engineering, as in mathematics, extreme cases are often illustrative); the optimum convoy size must consider the disadvantages of increasing size as well as the advantages. The obvious disadvantages are that the convoy can move no faster than its slowest ship, and that the arrival of a convoy swamps port facilities, greatly increasing turnaround time.

In fact, the study is guilty of suboptimization; what has been optimized is the skirmish between a convoy and a submarine pack, for which the measure of effectiveness is the ratio of submarines sunk to merchant ships sunk. What should have been optimized is the Battle of the Atlantic, for which the proper measure of effectiveness is the goods delivered to the eastern shore of that ocean. Of course, one can quibble about modifications of this measure depending on the length of the war (do we want to maximize goods delivered during the next month or during the next year?) and the desirability of saving human lives (at least on our side), but the principle is clear: if the convoy is too large, it will take so long to assemble, load, sail, unload, and return that the amount of goods delivered may be considerably less even though we lose less shipping.

But even this viewpoint is a suboptimization, because the real objective is not so much to win the Battle of the Atlantic as to win the war.

And when the German submarines went elsewhere, they indicated that we had gone too far. It is a principle of competitive situations (and, to the extent that they can be considered games against nature, all stochastic systems) that when we have achieved our optimum strategy, we are indifferent as to what the enemy (or nature) may do. This concept is made formal in game theory. It follows that if the enemy (or nature) has a clear-cut preference available, we are not at the optimum. In this case, if the German submarines could clearly do better by leaving the North Atlantic, we must have made our convoys too large.

But even this is a suboptimization, because the objective of winning the war should be subordinated to the objective of optimizing the postwar world. In this example, such considerations would be stretching things; but it was, for example, a serious question in the decision to drop the bomb on Nagasaki. And of course, there are even higher objectives. So what is the system designer to do if he is several echelons farther down (e.g., designing an antisubmarine guided-missile system aboard one of the escort vessels)?

In answer, Hitch suggests "the relevancy of economics" which "involves the analysis of relations between suboptimizations at lower and higher levels." We would add that, while absorption in the problems of higher levels can lead to paralysis or, worse, severe political repercussions, the designer should always be cognizant of their problems and the effects of his actions on them. Most important, he should know the level of his own sponsor and select appropriate criteria with his sponsor.

As a rule of thumb, in addition to his own level, a systems designer should think one level up and one level down. He should think one level up because the task as he receives it is not completely defined (for the reasons discussed earlier) in that a problem cannot be completely formulated before it is solved. Similar arguments dictate that the designer needs also to think one level down.

The Compleat Systems Engineer

Finally, we turn to the man who must do all of this—the systems engineer. The use of the word "systems" is sufficiently pervasive that every large organization invariably has some people who are identified as systems engineers. Many of these people do not have, and need not have, an understanding of the broad range of systems concepts as they have been presented here. These people may have a well-defined, static role to fulfill in their organization; and the role, important as it may be, may have little requirement for these concepts.

What we mean by the "compleat" systems engineer is the man who is the creator, the innovator, the synthesizer of systems. This is the man who needs to have the "big picture"; who must see the path from systems requirements to systems operation; who can make decisions, implement ideas and bring the system into being.

Clearly, this man must be, in some sense, a generalist rather than a specialist. The ideal systems engineer is a "T-shaped man," broad, but deep in one field. His depth is provided by scholarly experience—a Ph.D. or equivalent—and the breadth by extended interests and abilities. Frequently he must become a "6-month expert" in a new field, such as meteorology or television or electroencephalography, but he will find that his background in mathematics and engineering will enable him to learn enough in a short time to allow him to work with real experts in the field.

In addition to systems engineering, he must know a good deal about administrative and marketing matters. In particular he must be a good salesman, because regardless of the merit of his ideas, he must convince some sponsor that his project is more worthy of support than the numerous other proposals which are invariably competing for the limited financing, equipment, or time available. He must know about the project system of management; he must know about costs and accounting procedures; and he must know about organizational and administrative politics, which probably cannot be learned from any book.

Of course, the man who knows all of this does not exist. In practice, the systems engineer does not need to know everything. What he needs to know is everything that is pertinent to his particular problem. In that sense, there are thousands of systems engineers who come remarkably close.

BIBLIOGRAPHY

Baumgartner, John S., *Project Management,* Irwin, 1963.

Bellman, Richard, *Dynamic Programming,* Princeton University Press, 1957.

Bryson, Arthur E., and Yu-Chi Ho, *Applied Optimal Control,* Blaisdell, 1969.

Chestnut, Harold, *Systems Engineering Methods,* Wiley, 1967.

Chestnut, Harold, *Systems Engineering Tools,* Wiley, 1965.

Dantzig, George B., *Linear Programming and Extensions,* Princeton University Press, 1963.

Ford, L. R., and D. R. Fulkerson, *Flows in Networks,* Princeton University Press, 1962.

Goode, Harry H., and Robert E. Machol, *System Engineering,* McGraw-Hill 1957.

Hall, Arthur D., *A Methodology for Systems Engineering,* Van Nostrand, 1962.

Machol, Robert E. (Ed.), *System Engineering Handbook,* McGraw-Hill, 1965.

Machol, Robert E. (Ed.), *Information and Decision Processes,* McGraw-Hill, 1960.

Machol, Robert E., and Paul Gray (Ed.), *Recent Developments in Information and Decision Processes,* Macmillan, 1962.

Miller, Robert W., *Schedule, Cost, and Profit Control with PERT,* McGraw-Hill, 1963.

Schwarz, Ralph J., and Bernard Friedland, *Linear Systems,* McGraw-Hill 1965.

System Engineering Management Procedures. Air Force Systems Command Manual (AFSCM 375-5), United States Air Force, February 1964.

Wilde, D. J., and C. S. Beightler, *Foundations of Optimization,* Prentice-Hall, 1967.

4

Decision Analysis in Systems Engineering

RONALD A. HOWARD

Professor of Engineering-Economic Systems
School of Engineering
Professor of Management Science
Graduate School of Business
Stanford University

Ronald A. Howard is a Professor of Engineering-Economic Systems and a Professor of Management Science at Stanford University. He received his Ph.D. degree in electrical engineering from the Massachusetts Institute of Technology and served on the faculty until 1965. He is the author of *Dynamic Programming and Markov Processes* and *Dynamic Probabilistic Systems*. He has editorial responsibilities for *Management Science, Operations Research, IEEE Transactions on Systems, Man, and Cybernetics, Journal of Optimization and Control,* and *Journal of Socio-Economic Systems.* He is the Series Editor for the Wiley series, Decision and Control. Howard's research and consulting interests are currently concentrated on decision analysis.

The past decade has seen the development of a new profession—decision analysis, a profession concerned with providing a rational basis for decision-making. While it may seem strange that people can make their living by helping other people make decisions, that is just what decision analysts do. So that we can better see the need for this new profession, let us start by taking a look at the kind of decision-making we use in our everyday lives (see Fig. 1).

Descriptive Decision-making

In this descriptive view of decision-making, we first examine the environment of human decisions. The environment can be described by several characteristics. As we go through these characteristics, think of them not only in terms of modern corporate or governmental decisions, but in terms of any decision—personal, romantic, or historical. Perhaps we might even think back to the dawn of history when the caveman was trying to decide whether to take one path or another in order to avoid the saber-toothed tiger. He saw his environment, as we now see ours—uncertain. If there is any one attribute of the environment that gives us the most difficulty in decision-making, it is uncertainty. Furthermore, the environment is complex—we see many different factors interacting in ways we often cannot understand. It is dynamic. It evolves over time. What we do today has effects that may not be evident for years.

Unfortunately, in business, military, or national problems the environment is often competitive. There are hostile intelligences that are trying to make life better for themselves at our expense. Perhaps, most unfortunate of all, our resources are finite—in spite of the exhortations of religious leaders over the centuries, man perceives himself as being a limited creature who has to allocate what he has, rather than to expand it.

The typical human reaction to these characteristics of the environment is confusion or worry, whether it be corporate—and there are

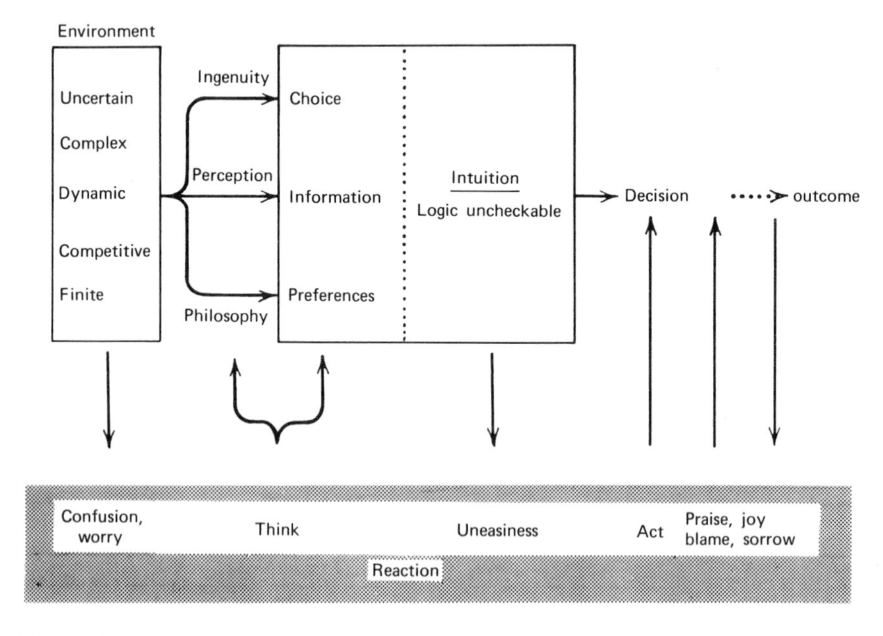

Figure 1 Decision-making (descriptive).

many worried corporate executives—or individual. But man has weapons available in his fight with the environment—his fight to make a decision. We separate these weapons into three types.

First, man has ingenuity. He uses that ingenuity to conceive and formulate different courses of action; that is, he has the potential of choice. Second, he has perception. He can learn from what he sees. He can form judgments about the world. In other words, he can gather information about his environment. Finally, and in some ways perhaps most important, he has a philosophy. He has guiding principles of his life that give him preferences among the various outcomes that he might obtain from his decisions.

Although we have only looked at the personal application of these ideas, corporations, colleges, and governments also have these attributes. We might combine them into the idea of thinking. This is the process of thought—the process that man brings to the problem of making a decision.

At this point let us define a decision. A decision is an allocation of resources that is revocable only at a cost in some resource, such as time or money. For practical purposes we should think of it as an irrevocable allocation of resources. It is not a mental commitment to changing the state of the world or to carrying out some course of action, but rather some actual physical change.

Intuition

The process by which man makes the overwhelming number of decisions in his life (or a corporation in its life) is intuition. He uses an intuitive process to balance the choices, the information, the preferences that he has expressed and to arrive at a course of action—to arrive at a resource allocation. We cannot say much about how intuition operates, but we have all met people who operate intuitively. Indeed, we all make intuitive decisions in our own lives every day: which route to take to get to work, when to get up in the morning, and so forth. We would be foolish to substitute any other principle for intuition in the majority of the decisions we make. But there are some decisions that we, as individuals or as organizations, face that are so important—so crucial to our existence, survival, and gathering of joy—that we must strive for a better way of making them.

The characteristic of intuition that is most bothersome to us is that its logic in uncheckable. If a person were the chairman of the board and made a decision by intuition, he might say, "Well fellows, I've read all the reports and, having thought it over, I think we ought to merge with Company X." While such a decision could be a great idea, we really have no way of evaluating it. There is no way of checking step-by-step to determine whether this decision is the logical consequence of the choices, information, and preferences that were available to the decision maker. It all went on in his mind—behind closed doors, so to speak. While a one-man company may be able to get away with such a decision, in our increasingly interdependent corporate world or in our society it becomes increasingly important for a man to be able to show people why he arrived at a particular decision. It is also important for them to be able to see what changes in factors surrounding that decision might have led to a different decision.

Thus we often find that one result of the intuitive decision-making process is uneasiness on the part of the individual or the organization making the decision. You would be surprised at the number of corporate decision makers who arrive at a decision on intuitive grounds and then, after the fact (after they have made the mental commitment, but before they have written the check), come looking for some better way of making the decision because they are uneasy about whether that decision is consistent with their choices, information, and preferences.

Decision Versus Outcome

One other thing we ought to mention about descriptive decision-making is the unfortunate human tendency to equate the quality of the decision with the quality of the outcome it produces. Each decision is followed by an outcome that is either joyful or sorrowful, for example: the surgeon decides to amputate the arm, and the patient either recovers or dies; the investor decides to buy some new stock, and the stock either makes money or loses money. We tend to say if the stock lost money, or if the patient died, that the decision maker made a bad decision.

Well, logically, that is indefensible, because the only way you can evaluate the quality of a decision is by whether it is consistent with the choices, information, and preferences of the decision maker. While we all prefer a joyful outcome to a sorrowful outcome, only the decision is under our direct control. We must seek aid in exploiting that control to the fullest extent, but we must distinguish the quality of the decision from the quality of the outcome.

Here is a simple illustration. Suppose that a person were offered the opportunity to call the toss of a coin for $100. If he calls that toss correctly he wins $100. If he does not, he wins nothing. There are very few people who wouldn't like to play such a game. Suppose that the offer were for a payment of $5. We can be sure that many people would like to play such a game for $5. Then, picture a line of people waiting for their turns. The first one comes up, we take his $5, we toss the coin, he calls it, and he loses. Now what? What do we say? We say he had a bad outcome. The next one comes up; he also pays his $5, and he wins. He had a good outcome. These people have both made the same decision. It was a good decision, but making that good decision is no guarantee of a good outcome. Speaking loosely, making a good decision is only doing the best we can to increase the chances of a good outcome.

One thing that anyone who deals with decision analysis should keep in mind is the importance of differentiating between the quality of the decision and the quality of the outcome. This distinction is the very beginning of the study of decision-making. It is this transcendence of the intimacies of outcome by conceptualizing the decision-making process that allows us to study formally what "good decision" means.

In most cases we do not really know what is a good decision. We are so used to characterizing the kind of decision that was made by the kind of outcome produced that we really have not until now had a procedure—an engineering analysis, a science, if you like—for recognizing a good decision. One of the "reasons to be" for decision analysis is to formulate the idea of what a good decision is, and to formulate it in quantitative terms that can be conveyed from one person to another, compared from one situation to another.

So much for descriptive decision-making. We will probably be using it a lot in our personal lives and in our organizations' lives, but it has the shortcomings of intuition that we think we can now transcend.

Decision Analysis

In this chapter we examine an alternative to descriptive decision-making, an alternative called "decision analysis." Here is a very brief definition. It is the balancing of the factors that influence a decision and, if we wanted to add another word, a *logical* balancing of the factors that influence a decision. Typically these factors might be technical, economic, environmental, or competitive; but they could also be legal or medical or any other kind of factor that affects whether the decision is a good one.

The Precursors of Decision Analysis

Decision analysis is a term we also use to describe the outgrowth of two earlier fields, namely, decision theory and systems modeling methodology. Decision theory was largely the province of academics until very recently. They treated the question of how to be rational in very simple, but uncertain, situations dealing with balls in urns, coin-tossing, small amounts of money, and the like. But it turned out that there was enough meat to the question of what is a good decision (even in simple cases) that theorists for years—going back to Bernoulli in 1738—have been worried about what really constitutes a good decision. However, decision theory was a theory for very simple decisions and certainly far from application to the complex corporate or even personal decisions we face today.

Over the past 30 years, we have also seen the development of a systems modeling methodology. That systems modeling methodology provided means of treating the complex and dynamic aspects of the environment in a way that had never been contemplated before. Of course, the advent of the computer played a large role. Decision analysis is the child of both of these developments. It is a way to combine the ability to handle complexity and dynamics with the ability to handle decision-making in the face of uncertainty into a single discipline that can treat all three simultaneously.

A Language and a Procedure

This new discipline has two interesting aspects. First, it is a language and philosophy for decision-making. It is a way to talk about the decision-making process even if you never set pencil to paper to do a computation. Indeed, organizations that have begun to think in this way—to use this language and philosophy—can never, it appears, revert to their old ways of thinking. It is a kind of reverse Gresham's law: the specification of language and clarity of concept keeps us from thinking about decisions in ways that might not be fruitful in the making of them.

But more than that, as far as the profession itself is concerned, is the idea that decision analysis is a logical and quantitative procedure. It is not simply a way to talk about decision-making; it is actually a way to make a decision. It is a way to build a model of a decision that permits the same kind of checking and testing and elimination of bugs that we use in the engineering of an automobile or an airplane. If it were not for the fact that "decision engineering" somehow implies a kind of manipulation of the decision-making process rather than an analysis of it, this field might be called decision engineering rather than decision analysis.

The Decision Analysis Formalism

How does decision analysis differ from intuitive decision-making? In some ways, not at all; in other ways, very significantly. First, consider the environment (see Fig. 2). There is "bad news" on that score because the environment is still uncertain, complex, dynamic, competitive, and finite. We shall have to live with it—decision analysis is not a "crystal ball" procedure, much as people wish it were. So we will still be confused and worried when we start out on a decision problem. Furthermore, there is no hope for people who do not like to think, because we must be ingenious, perceptive, and philosophical in order to carry out decision analysis. So it is not much help yet.

Choice. Where we start to get help is now. First, let us go through the three aspects of ingenuity, perception, and philosophy, one by one, and see how they are treated within this new discipline. The idea of choice is spelled out by enumerating specific alternatives that are available in this decision problem. They may be finite alternatives, like amputate or do not amputate; or they may be alternatives described by continuous variables, such as the capacity of a plant, the price of a new product, or even the size of a budget that will be set for

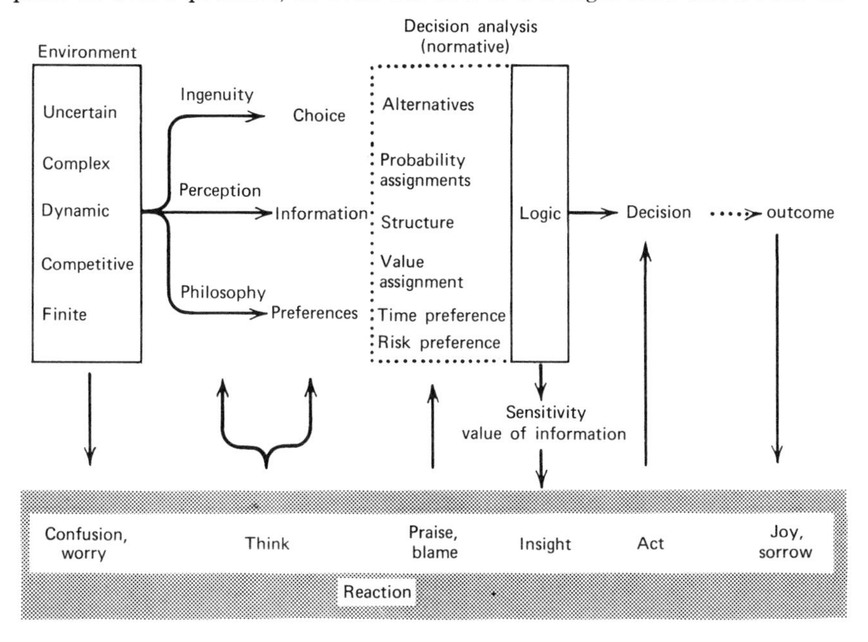

Figure 2 Decision-making using decision analysis.

a lower-level organization. The point is that alternatives are described quantitatively and not in general terms.

Incidentally, the process for developing new alternatives is one that we wish we were more able to comment upon. Ingenuity is required; there is just as much need for creativity in the process, as analyzed here, as there is in intuitive decision-making. The only advantage that decision analysis can bring to the search for new choices or alternatives is the same kind of help that any analytic model provides when brought to bear on a problem. For example, using engineering models, while not directly synthetic in most cases, can lead to insights into existing designs that suggest new alternatives, new ways of solving the engineering problem. The same thing can happen here, but it is not part of the formal structure.

The Encoding of Information. The first new thing comes about when we look at information. We represent information in two ways. We characterize uncertainty by means of probability assignments, and we represent relationships by means of models, that is, by structuring the problem. Let us talk about structuring first.

STRUCTURING. Structuring is the kind of "head bone connected to the neck bone" arrangement we find in physical models. But now we are talking about a decision model: a way of representing the underlying logical relationships of a decision problem—be it national, legal, industrial, or whatever—in a mathematical model that shows what affects what. This is something that is discussed in other parts of this book—the process of modelling and using computers in modelling. Although it is not customary to build formal models of decisions in the same way that we build formal models of other engineered systems, decision analysts do just that. In fact, anyone who is going to be a professional in this area is required to be conversant with modern modelling techniques.

Now let us return to the treatment of uncertainty.

PROBABILITY. Since many readers may not be familiar with the field of decision analysis, there is no reason to examine the long arguments that used to go on as to whether probability was a state of mind or a state of things. Decision analysts believe that it is a state of mind, a way of representing one person's uncertainty about a particular event or variable and that it has no necessary interpretation whatever in terms of real-world long-run frequencies.

The whole idea of describing uncertainty by means of probability assignments has come about only in the last ten or twenty years. Before that, probability was thought of as the province of statisticians—a region that only experts could enter. We may still need an expert to do more complex probabilistic manipulations, but we do not need an expert to think in probabilistic terms, which is what decision analysis requires. In some ways this is the most unique part of the decision analyst's trade, that he is able to deal effectively with the assessment and implications of uncertainty.

Lest the reader think this aspect is being overemphasized, here is a quote from a book that is in an entirely different field. It is a non-technical book called *The Search for Authenticity* by James F. T. Bugental who is a psychoanalyst. He writes:

> Let us pause to examine this quest for certainty. By "certainty" I mean the opposite of contingency. Having survived a disastrous fire in our neighborhood and being concerned about my home, I decide to investigate the likelihood I would not be so fortunate again. I find my odds are 1000 to 1 against the likelihood that my house will burn, but I am not content and so have the brush cleared back some distance. Now the odds are 1500 to 1, I find. Still concerned, I have an automatic sprinkling system installed. Now I'm told my odds are 3500 to 1. However I try, though, I must recognize always that I cannot achieve certainty that the house will not burn. I may do much, but I can't be sure that a nuclear firestorm will not make my efforts vain. I may build my house underground but still I can't be sure but that the earth might be drawn closer to the sun and the whole world thus be ignited.
>
> Now these are ridiculous extremes, of course, but the point remains: there is no true certainty to be had. So it is with any issue. Nevertheless, we seek that certainty constantly. We buy insurance, seat belts, medicines, door locks, education, and much else to try to protect ourselves against tragedy, to secure good outcomes. So long as we recognize we are dealing in probabilities, such choices can be useful. But every therapist has seen the pathology of seeking for certainty instead of better probabilities.[1]

One could take that last sentence, replace "therapist" by "decision analyst" and say that every decision analyst has seen the pathology of seeking for certainty instead of better probabilities. Every corporation the author has ever encountered believes that the secret of security comes from making things certain. In a course at Stanford University where students actually go out and do decision analysis as part of the course work, one of the presentations concerned the idea of "proven" reserves of mining ore. There we see the intent to make something certain when it is not certain, because a proven reserve is not a proven

[1]James F. T. Bugental, *The Search for Authenticity,* Holt, Rinehart and Winston, 1965, pp. 74–75.

reserve—it is something with probabilities attached to it, probabilities in terms of the amount, the type, the cost of extraction, and so forth.

Another example from class concerned how to treat people who have angina. Do you give them special new surgical procedures or do you give them conventional medication? Here the point came up again that there were some procedures that were "proven" medical treatments and others that were unproven. In other words, the world had again been divided into things that were OK and certain and approved by society, and others that were not OK, and there was a very small area in between. In these examples, as we all see in our own lives, we are continually trying to get the uncertainty out of the way because it is so painful with which to deal (as Bugental, the psychoanalyst, says).

We see the same thing in corporations. They attempt to set budgets, goals, and growth rates in an endeavor to ascertain what is basically uncertain. I claim that the way to corporate health is not to try to make the world certain, but to live with it in its present uncertain state, to act in the best possible way given the kind of world we live in. Bugental also sees that as the key to mental health, so I guess we agree even though we are in different fields.

We'll return later (see the question period at the end of this chapter) to how we go about making probability assignments. For the moment, let us just say that there is a way to do it and that such assignments become one of the two parts of the total encoding of information (the other being structure) that lead finally to putting into the model what we know.

So if we had to characterize the inputs to decision analysis, we would say choice is what we can do, and information is what we know. Now we come to the third: preferences—what we want.

The Establishment of Preferences. It turns out that because of our previous inability, or perhaps a better word would be reluctance, to deal with uncertainty, we have never gotten in most decision problems to the question of what we really want. It is a very interesting exercise to take a guy who has a tough decision because there is a lot of uncertainty in it, and ask him, "Well, suppose I eliminated all the uncertainty, suppose I told you for sure what was going to happen here and here and here, then, what would you like?" He often does not know. Think about it in terms of new possible states-of-being for the United States. If we could snap our fingers and have any state we wanted, which would we want?

Decision analysis separates uncertainty from the establishment of preferences. Once we tell a decision maker, "Look, let me worry about the uncertainty, that's my business." "We just encode that and do the best thing we can with the uncertainty that exists, given the structure and alternatives." "What do you really want?" "How much more is this outcome worth than that?" Then he has a problem he can work on constructively.

We break the idea of preferences up into three categories: The first kind of preference we call value assignment; the second, time preference; and the third, risk preference. Value assignment is concerned with situations where you have different outcomes and you say, "How much more is this outcome worth than that?" "What's the relative value of the two?"

VALUE ASSIGNMENT. Here is an example that may drive this topic home. Suppose we consider a medical case with which we can all identify. You walk into the doctor's office and he says, "You've got acute something-or-other and we're going to have to do this to you in the hospital, and you're going to be there for a day with severe pain, and then you'll be all right." And you say, "Well, what's severe pain?" And he says, "It's like pulling a wisdom tooth without an anesthetic." Each of us has his own opinion on whether this is a suitable torment of Hell, but at least you can think about what that outcome would mean to you—how joyful or sorrowful it would be. We have to allow for all tastes.

However, we have another alternative: to take a magical drug that will produce an instant painless cure for your malady. You have a choice—either a day in the hospital with pain and then cure, or the instant cure with the magical drug with no side effects (see Fig. 3). How much more would you pay for the instant cure via the magical drug compared to a day in the hospital with pain? Magical drugs are expensive, so let us see how much you would pay in addition to what the hospital trip would cost in order to obtain the drug. Would you pay a dollar? Sure, you'd pay a dollar. $10? Sure, that wisdom tooth is

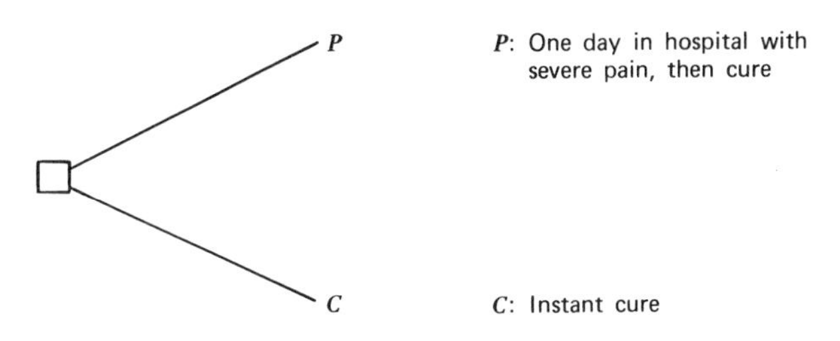

Figure 3 Value assignment.

pretty painful. How about $10,000? Your reply would probably be: "You must think I'm made of money." "Where am I going to get $10,000?"

So now we are bounding you. We may have a few millionaire readers who laugh at $10,000; we may have struggling students who say, "Give me ten bucks and take out all four wisdom teeth for all I care." But each of us in his own financial situation can say how much more he would pay for one than the other. Notice that we are making a decision out of the value question. We are saying, if you could go down each route, how much more would we have to make the instant cure cost before you would be indifferent. So one of the key ideas is to use the idea of comparison and adding positive increments to one side or the other until you say, "O.K., I cannot tell the difference." That is the value question. Given an outcome that occurs now with no uncertainty, how much do you like it?

TIME PREFERENCE. The next question we face is time preference. Time preference concerns the worth we place on values that are distributed over time. This involves what we call the "greed-impatience trade-off." We are usually willing to accept less if we can get it sooner. Establishing the time preference of an individual or a corporation is not simple, but we can demonstrate that it is very important.

One case we worked on involved a person who had to choose between having an operation for a kidney transplant or being put on dialysis indefinitely with a kidney machine. It turned out that the whole question boiled down to time preference for him and, on further investigation, it developed that what was important to him was to live until his children got through college. So his time preference had an interesting structure. He placed a high value on being alive until some point x years in the future, and after that not so much. That is an unusual kind of time preference, but one any complete theory has to be able to accommodate.

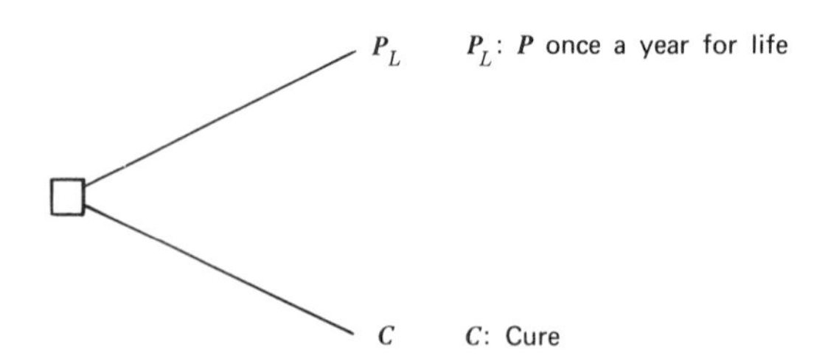

Figure 4 Time preference.

How can we demonstrate time preference in the medical example? In the medical case, suppose we think about a new event P_L, which is getting that one day in the hospital with pain once a year for life, as opposed to an instant cure now (see Fig. 4). Every year you have to go back to the hospital for one day and undergo the painful treatment or, on the other hand, you can get an immediate cure. It is clear that the instant cure is now worth more than it was before, since you were already going to be in for the first year anyway, but how much more depends on your attitude toward the future. If you say that anything that happens more than 30 days out you really do not care about, you will have one answer to your question. But, if you are very much concerned about retirement income, you will have another attitude. The way you answer this question will give us a lot of insight into how you value time.

Of course companies and nations face time preference questions. If we are thinking about setting up a national park system or other long-run investment, we are going to have to think about how much benefits in the future are worth relative to benefits today.

RISK PREFERENCE. Although value and time are certainly important, probably the most unusual aspect of the profession we are talking about is its ability to handle the third of the three, namely, risk preference. It is easy to demonstrate risk preference. Risk preference is the term we use to describe the fact that people are not expected-value decision makers; that is, they are not willing to choose among alternatives simply by comparing their expectations. (The expectation of an alternative is computed by multiplying each monetary prize by the probability of receiving it and then summing the products for all prizes.)

Suppose we said, "How many of us would flip a coin, double or nothing, for next year's income?" Whatever we would have gotten next year, we will either get twice as much, if we call the coin correctly, or nothing, if we call it incorrectly. Now that situation has an expected value equal to next year's salary, and anyone who is willing to make a decision on the basis of expected value should be marginally willing to engage in such a proposition. It is doubtful that we would have many takers, because there is nothing in it for a person. Suppose we pay each person a dollar to play; so now there is something in it for the taker. But most of us still would not do it. What if we say, "All right, what fraction of next year's salary would we have to pay to induce a person to engage in this gamble?" If we pay the fraction 1, then the taker has nothing to lose. He will get next year's salary anyway, and everyone will try it. If we pay the fraction 0, no one will try it. The real question is what fraction of next year's salary do we need to offer? Typically, numbers like 60–95 percent might be appropriate, 95 percent corresponding to the person who has substantial financial commitments and just does not see how he is going to make it, whereas the smaller fraction would correspond perhaps to the footloose bachelor who figures he can always go and live on the beach if he does not get any money next year. So it will be very specific to the person, to his own environment, his own tastes; and, in that sense, everything we are talking about is unique to the individual. It is appropriate to the decision maker and is not for the public at large.

To extend our medical discussion to this case, all we have to do is think about an imperfect magic drug. Unfortunately, the magic drug that might cure us will, now, also be able to kill us. So we will have to choose between going to the hospital with the day of pain or taking the magic drug, now costless, but which will kill us with probability p and cure us with probability $1-p$ with no pain—no side effects (see Fig. 5). The question is what is the probability p such that we are indifferent toward the day in the hospital with pain and taking the magic drug. Think about it; imagine being placed in this situation. It is not a very unrealistic situation; there are cases just like this that occur in medical practice.

What if $p = 1/1000$? One chance in a thousand we are going to die from the drug versus a day in the hospital with pain. The answer would probably be: "Dying's pretty bad, I don't like that." What about one chance in ten thousand? A typical reaction: "I'm feeling pretty lucky today—one in ten thousand, I might just do that." The point is, once we establish the value a person places on his life—which is another long story—and the value of a cure relative to a day in the

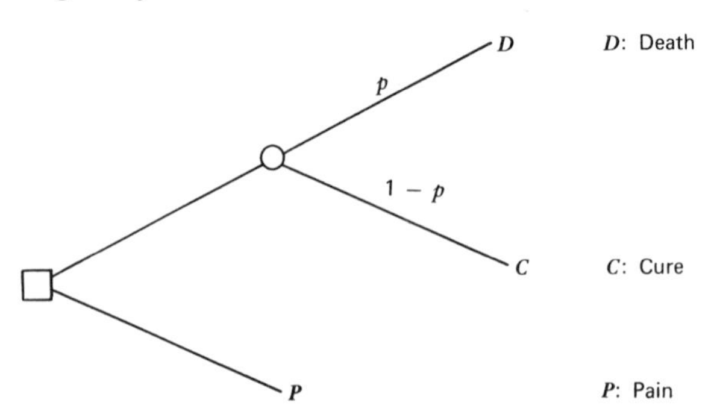

Figure 5 Risk preference.

hospital with pain, then the number p would certainly give us insight into their attitude toward risk and would allow us indeed to start building a description of their risk preference.

The Logical Decision. When this has been done, when we have carried out this procedure and have established preferences, the values placed on outcomes, the attitude toward time, the attitude toward risk (and there is a methodology for doing all of this), when we have established the models necessary for the decision one is making and have assessed probabilities as required on the uncertain variables, then we need nothing but logic to arrive at a decision. And a good decision is now very simply defined as the decision that is logically implied by the choices, information, and preferences that we have expressed. There is no ambiguity from that point on—there is only one logical decision.

This allows us to begin to assign praise or blame to the process of making the decision rather than to the ultimate outcome. We can do an analysis of the decision and make sure it is a high quality decision before we learn whether or not it produced a good outcome. This gives us many opportunities. It gives us the opportunity to revise the analysis—to look for weak spots in it—in other words, to tinker with it in the same way we can tinker with an engineering model of any other process.

The Value of Information

If this were all decision analysis did, it would be impressive enough, but from it we also get other benefits. We obtain sensitivities to the various features of the problem and we learn something that I think is unique to decision analysis called the "value of information." The value of information is what it would be worth to resolve uncertainty once and for all on one or more of the variables of the problem. In other words, suppose we are uncertain about something and do not know what to do. We postulate a person called a "clairvoyant." The clairvoyant is competent and truthful. He will tell us what is going to happen—for a price. The question is what should that price be. What can we afford to pay to eliminate uncertainty for the purpose of making this decision?

Of course we do not have real clairvoyants in the world—at least not very often—but the clairvoyant plays the same role in decision analysis as does the Carnot engine in thermodynamics. It is not the fact that we can or cannot make it, but that it serves as a bench mark for any other practical procedure against which it is compared. So the value of clairvoyance on any uncertainty represents an upper bound on what any information-gathering process that offers to shed light on the uncertainty might be worth.

For example, if we find in the medical problem that the value of clairvoyance on whether or not we are going to die from the drug is $500, then that means that we should not pay more than $500 for any literature search or anything else that would provide only imperfect information with respect to whether or not we are going to have this problem.

That is a revelation in itself to many people—the fact that one can establish a hard dollars and cents number on the value of information to us in making a decision, and hence can use that number to guide what information-gathering processes we might participate in.

The Medical Problem Evaluated. It is hard to demonstrate very simply how to do such a calculation, but let us try by taking the medical example and putting some numbers in it (see Fig. 6). The patient has the choice of taking the magic medicine or not. If he does not take it, then he is going to get the pain; we will consider that as a reference point of value $0. If he does take the medicine, let us suppose he has one chance in a thousand of dying and 999 in a thousand of getting the instant cure. We have also put in numbers here saying that the cure is worth $100 more than the pain. He is a relatively poor person, but he would pay $100 more for the painless cure than he would for spending a painful day in the hospital. Now for death—what is the value of life to a person? This person has set the value of his life at $100,000.

Notice that we "set" the value of his life. What is meant by this is that he wants the designers of public highway systems and airplanes to use the number $100,000 in valuing his life. Why does he not make it a million dollars? If he does, he will have more expensive rides in airplanes, more expensive automobiles, and so forth. He does not get something for nothing. If he makes it too small, he had better be

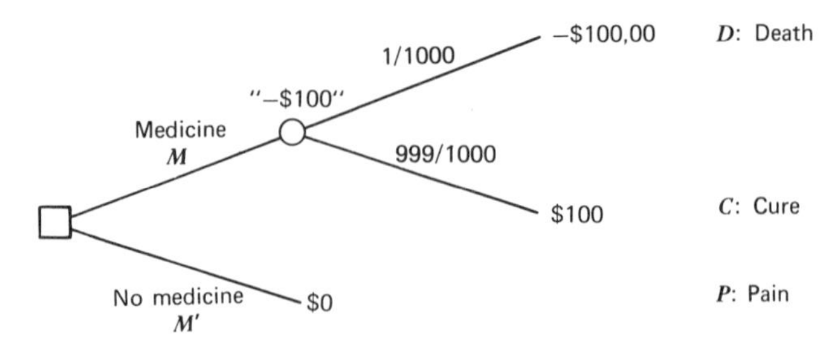

Figure 6 The medical decision.

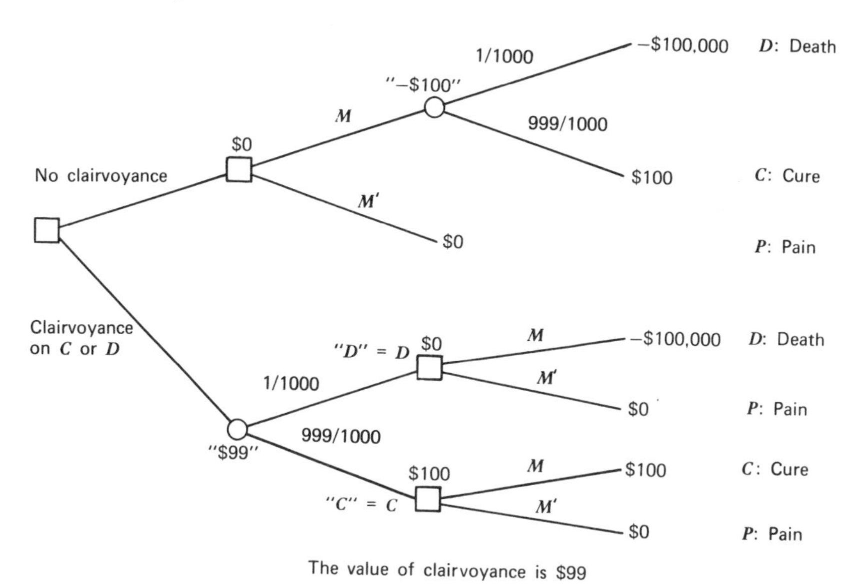

Figure 7 Value of clairvoyance computation.

wearing a helmet every time he enters his car. So it is a decision for him as to what number he wants the decision makers to use in this completely logical world that we are talking about.

The number –$100 in quotes (in Fig. 6) means that our patient has said that one chance in a thousand of losing $100,000 and 999 chances in a thousand of winning $100 has a value to him of –$100. In other words, we have to pay him $100 to get him to take on this uncertain proposition. It is clear that, comparing –$100 to $0, he is better off deciding not to take the medicine. So for him the probabilities, values, and attitude toward risk leading to the –$100 assessment of this whole uncertain proposition, the best decision is to forget about the medicine.

Clairvoyance. Now the clairvoyant arrives. If the individual we are talking about does not patronize the clairvoyant, then he does not take the medicine and makes nothing. If, on the other hand, he does buy the clairvoyance on the question of whether death will occur, what will happen? First, the clairvoyant will tell him whether he is going to die if he takes the medicine (see Fig. 7). We have "*D*" in quotes here, meaning that the clairvoyant says he is going to die, equivalent to his actually dying because the clairvoyant is truly prophetic. "*C*" means the clairvoyant says he is going to be cured. Since the probability the clairvoyant will say he is going to die has to be the same as the probability that he really will die, he has to assign one chance in a thousand to getting that report from the clairvoyant. Now suppose the clairvoyant says he is going to die. Obviously, he ought not to take the medicine in that case, and he will make nothing. If the clairvoyant says he is going to be cured without dying, then he is better off taking the medicine, and he will make $100. Since the payoff from the clairvoyant's saying that he is going to die is $0 and from not going to die is $100, and since there are 999 chances out of a 1000 that the clairvoyant will say he is not going to die, just by looking at that lottery we can see it will be worth almost $100 to him. He has 999 chances out of a 1000 of winning $100, and only one chance in 1000 in winning $0.

Let us suppose he evaluates the whole uncertain proposition at $99. If he does not buy the clairvoyance, he is looking at $0; if he does buy it, he is looking at a proposition that is worth about $99 to him. Thus, the value of the clairvoyance would be $99.

So here is an uncertain proposition with all kinds of big numbers running around in it, yet a very simple calculation based on his attitudes toward risk, life, death, and pain says he should not be willing to pay more than $99 to know for sure whether he would get the unfortunate event of death if he should take the drug.

Similarly, in any other decision problem—and there are some very, very complicated ones, involving many jointly-related variables—we can establish an upper bound on the value of information-gathering on any aspect of that problem. We can subsequently determine the best information-gathering strategy to precede the actual making of the decision.

The Decision Analysis Cycle

Let us begin with a word on methodology and then go on to an example. When doing a decision analysis it helps to organize your thoughts along the following lines. First, constructing a deterministic model of the problem and then measuring the sensitivity to each of the problem variables will reveal which uncertainties are important. Next, assessing probabilities on these uncertainties and establishing risk preference will determine the best decision. Finally, performing a value of clairvoyance analysis allows us to evaluate getting information on each of the uncertainties in the problem. The problem could be very complicated, involving many variables and months of modelling and analysis, but the basic logic is the same. The phases are: deterministic to evaluate sensitivities, probabilistic to find the best decision, and informational to determine in what direction new information would be most valuable. Of course you can repeat the process as many times as is economically valuable.

That is just to give an idea of how one does a professional decision analysis. Let us now turn to a case history to demonstrate the kind of problem that can be attacked in this way. Everything said so far has a naive ring to it. We can talk about betting on next year's salary, but we are really interested in not just the theory of decision analysis, but the practice of it.

A Power System Expansion Decision

Let us take an example from the public area. It concerns the planning of the electrical system of Mexico and is one of the largest decision analyses that has been done. It has been chosen because it comes closest to a problem in systems engineering. The specific question posed was: Should the Mexican electrical system install a nuclear plant and, if so, what should its policy toward nuclear plants in general be? Of course, we can not really answer that question without deciding how they are going to expand, operate, and price their system over time from here on out. So the real question is how to run the electrical system of Mexico for the rest of the century (see Fig. 8).

The Mexican electrical system is nationalized and very large—the size of several United States state-sized electrical systems. Because it is a complete national system, its planners have unique problems and also unique opportunities. The basic idea in working this problem was to look first of all at the various environmental factors that might influence the decision and then to look at the various measures of value that would result from particular methods of operation.

The Inputs

First, let us discuss the inputs. There are four input models: financial, energy, technology, and market. The financial models are concerned with the financial environment of the Mexican electrical system both in the world and the Mexican financial market. The inputs that these models provide are the amounts of money and the rates at which money can be borrowed from that source over time, with uncertainty if necessary. An input to this model is something called x which is picked up from the lower right. It is the book profit of the system. There is a feedback between the profitability of the system over time and the amount that it can borrow to support future

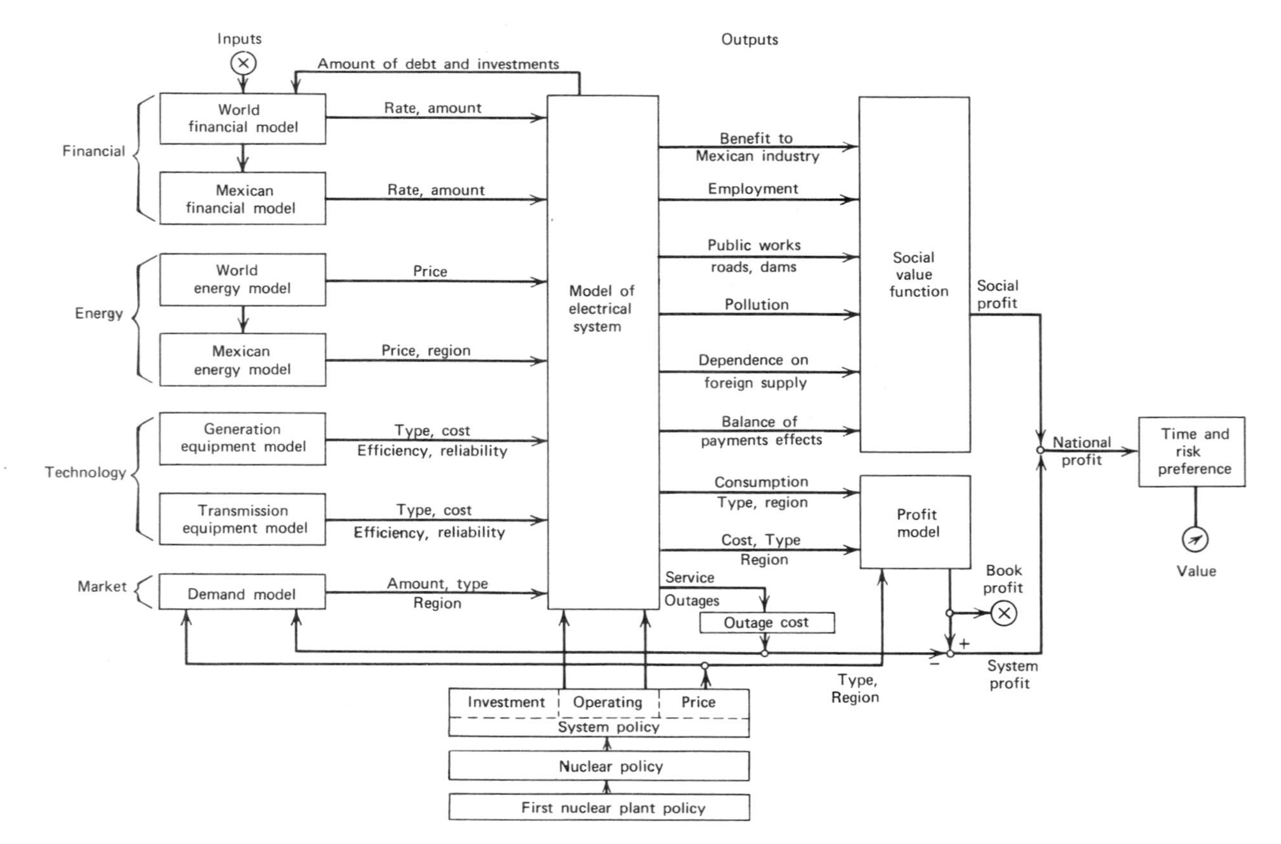

Figure 8 A decision analysis model of the Mexican electrical system.

expansion. The current amounts of debt and investment are also fed back.

The second type of input is energy costs, both in the world market and in the Mexican market. The interesting thing about Mexico is that it has just about every type of energy available: coal, oil, uranium, and thermal fields, and, of course, there are world markets in uranium and oil, at least, whose price movements over time would influence the economics of the Mexican system.

Next comes technology. This model describes generation and transmission equipment according to type, cost, efficiency, reliability. It includes such features as the advent of better reactors in the future and the possibilities of new and improved transmission systems which might make some of their remote hydro locations more desirable.

The last input model is the demand or market model, indicating by type and region the amount of electricity that would be consumed, given a pricing policy and given a quality of service. So these are the inputs to the model of the Mexican electrical system, which can then be run.

The Outputs

We will not go into the details of the rather sophisticated model which was prepared to describe operation and expansion of the Mexican electrical system. Of more interest in this discussion is the kind of outputs that were produced. There were the very logical ones of the consumption of electricity and the cost of producing the electricity by region to give a profit for the electrical system. This profit was what might be called the operating profit or book profit of the system, and is what the investor would see if he looked at the books of the Mexican electrical system. One modification to that profit which was considered was an economic penalty for system outages. A measure of the service provided by the system is added to the book profit to give something called system profit—which the investor does not see, but which the designer of the system does see. This penalty makes him unwilling to make a system that has outages for hours at a time, even though it might be more profitable if he looked only at the book profit.

The Social Value Function. But what is unusual about the outputs here is that many of them do not appear on the balance sheet of the corporation at all, but are what we might call social outputs; they enter into something called the social value function.

The decision maker in this case was the head of the Mexican electrical system. He felt many pressures on his position—not just the regular financial pressures of operating an electrical system, but social pressures coming about from the fact that his is a nationalized industry. For example, one of the things that was of concern to him was the benefit to Mexican industry. What would be the Mexican manufactured component of any system that might be installed? Another one was employment. How many Mexicans would be employed at what level if they went one route as opposed to another? Now we can see that the way we design the system is going to have major impacts on these kinds of outputs. If we have a nuclear system, then we might provide training for a few high-level technicians, but most of the components would be manufactured abroad; we do not have the army of Mexican laborers that we would if we built a hydro system in a remote location.

Another side effect is the public works that are produced by the generation choice. For example, with hydro you have roads and dams—that is access, flood control, and so on, that we would not have if we installed a large nuclear plant in the central valley of Mexico. Balance of payments is still another consideration. Mexico at that time had not devalued its currency; the currency was artificially pegged with respect to the free world rate. The question is, if we are going to have an import quota system to try to maintain this kind of disparity in the price of money, should we include that mechanism within the model or should we say other parts of the government are going to be responsible for making such adjustments. That is what the balance of payment effect is all about.

There are two outputs left that illustrate two different points. One is called dependence on foreign supply. At the time that this study started, there was a worry in the minds of the Mexicans that a nation supplying nuclear equipment might become hostile for some political reason and cut off the supply of repair parts, fuel, or maintenance facilities, much as the United States did with respect to Cuba. If that happened, of course Mexico would be in trouble. The question was, would this have a major effect on the decision, or would it not. They could buy insurance against it by stock-piling uranium until such time as they were able to establish alternate sources of supply. But it was a real worry, because they wanted to make sure they would be protected against any politically generated stoppage of equipment or supplies. By the end of the study, this whole area was of much less importance.

The other output was pollution. Originally the decision makers were not too interested in pollution. They said they could not afford to worry about it. And yet, if you have visited Mexico City, you know that atmospheric polution is very high. By the time this study was over, about one year later, they were very glad that they had provided a place in the model for pollution because they were now getting the same kind of citizen complaint that we get in the United States. Some of the things they were planning, like giant coal plants in the middle of Mexico City, were not acceptable any more.

The social outputs from the operating model entered the social value function to produce what we call "social profit." It represents social effects that do not appear on the balance sheet of the electrical system, per se. Social profit is combined with the system profit to produce national profit. Time and risk preference are expressed on national profit to give an evaluation of the system as a whole.

The problem that remained was to find a way to expand the Mexican electrical system that would produce the highest overall evaluation. Various optimization procedures were used to suggest installations of different types (gas turbines, nuclear, conventional, and hydro plants) to achieve this objective over the rest of the century.

The Nature of Policy

Let us briefly examine the question of what a policy for expansion of such a system means. A common policy in the past had been to establish a so-called plant list, which was a list of when each type of plant would be installed—in 1979 we are going to have an X-type plant in location Y. That is a little bit like asking a new father, "When is your son going to wear size-ten pants?" He could look at projected growth charts and say, "Well, I think it will be when he is nine years old." Another way to answer the question is to say, "Well, I will buy him size-ten pants when his measurements get into such and such a region." This is what we might call a closed-loop policy because we cannot say in advance when we are going to do it, but we have built a rule that will tell us the right time to do it.

So when we ask how is the system going to be expanded from here on out, no one can tell us: They can show us expected times for different things to happen, but indeed, only the program can determine what the effect on expansion of the future evolution of the system's environment will be. It has what we might call a self-healing property. If we foul it up by forcing it to put in a giant plant that it cannot immediately assimilate, then it is self-healing in the sense that it will delay and adjust the sizes and types of future plants until it gets back on the optimum track again. As a matter of fact, it is so much self-healing that it is hard to foul it up very much no matter what we do, because in the course of time it is a growing system that finds a way to get around any of our idiocies. In actuality, when they compared what this optimization system was doing with the designs produced by their conventional techniques using the same information, this system yielded superior results in every case.

The size of the Mexican study is interesting. It took approximately eight man-years, and was completed in one calendar year by a staff of decision analysts from the Stanford Research Institute Decision Analysis Group, plus four representatives of the Mexican Electricity Commission who were very competent in nuclear engineering and power system design. The programs and analyses are now being used in Mexico for continued planning of system expansion.

Other Applications

Other applications include industrial projects—should companies merge, should they bring out a new product, or should they bring a mine into production? All of these things are what we might call fairly conventional decision analyses by the criteria that we in the profession use.

Some interesting decision analyses have been done in the medical area, such as one recently performed on the treatment of pleural effusion, that is, water in the cavity between the lung and the chest wall. This was a one-year study done by a graduate student who, as far as the doctor (who was the lung expert) is concerned, completely encoded everything the doctor knew about pleural effusion. Later the doctor was asked if he developed this symptom would he prefer to be treated by this large decision model or by one of his colleagues. He said, without hesitation, he would rather use the model.

Another study that has just recently been completed is whether to seed a hurricane threatening the coast of the United States. It was based on a large experiment a few years ago on hurricane "Debbie" which indicated, but certainly not conclusively, that seeding a hurricane with silver iodide crystals would cause the wind to diminish about 15 percent. This in turn would lead to something like a 50 percent decrease in damage. The question now is—if you are the decision maker in the White House and here comes a big one, hurricane "Zazie," headed right for Miami—what do you do? Should you send the planes out to seed it knowing that, even so, there is a chance that it might get worse just because of natural causes and wipe out two cities instead of one? Or should you sit on your hands and possibly watch people get killed and property destroyed when they might have been saved? There is a tough problem. It has severe social impacts and is definitely a decision under uncertainty. Study of this problem was presented very recently to the President's Scientific Advisory Committee. They have formed a subpanel to see whether the conclusions should be put into effect.

Conclusion

We have tried to characterize what is a new profession—a profession that brings to the making of decisions the same kind of engineering concern and competence applied to other engineering questions. It seems fair to say that the profession has now come of age. We are able to work on virtually any decision where there is a decision maker who is worried about making that decision, regardless of the context in which it may arise. The only proviso is that the resources that he is allocating must be real world resources. We are not competent to allocate prayer because we can not get our hands on it—or love, which is infinite. But when it comes down to allocating money, or time, or anything else that a person or organization might have to allocate, this logic has a lot to be said for it. And indeed, as we have seen, the key is the idea of separating the good decision from the good outcome. Once we have done that then we have the same ability to analyze, to measure, to compare that gives strength to any other engineering discipline.

Question Period

QUESTION. Is the professional decision maker the man who is right out in the forefront making the decisions in his own name, or will there be a professional decision analyst who is like the ghost writer standing behind the man, the president, the corporate executive?

ANSWER. That is a good question. In the legal profession there is a maxim that the lawyer who defends himself in court has a fool for a client. And I think the same is true of decision analysis. I know that I would never want to be my own decision analyst because I am not detached. I want the answer to come out certain ways, subconsciously. For example, if I want to make a case for why I should buy a new stereo system, I will work like a dog to make sure that I have lots of variables in the analysis indicating that I am going to use it a lot, it is going to be very valuable to me, and it is not going to cost much. But when I bring in one of my friends who is a decision analyst, he will say, "Just wait a second." "How many days are there in the month?" "How many hours in a day?" "How often are you going to listen?" And pretty soon he has it down to size where I can say, "Yes, I am kidding myself, it just will not all fit together." So I think we will never get to the stage—nor should we—where the decision maker is the decision analyst. I think these are two very different roles and one can subvert the other.

What has turned out, however, is that some of the presidents of corporations who have been exposed to this kind of thing have begun to think the way that I am indicating here and to do very simple analyses on their own. And that is great. Everyone should know a little science, a little auto mechanics, and a little of everything, so they are able to do simple problems relatively well. But they should also realize that when they have a tough problem—one involving complexity, dynamics, modeling, and all the other things we have examined—then it is really time for a professional. We might take a medical analogy again. Most of the times when a person has a headache, aspirin is alright, but every once in a while it is a brain tumor and it is better not to take the aspirin. The important thing is to know the difference.

QUESTION. What about the systems analyst versus the decision analyst?

ANSWER. I see the decision analyst as the person who combines the complexity and the dynamic aspects of systems analysis with the ability to treat uncertainty and to measure preference—activities that are usually foreign to systems analysis. One of the problems with the systems field is that systems analysis is a much misunderstood term. Many groups and stakeholders in the systems professions have entirely different attitudes about what a systems analyst is. It can be everything from someone who riffles punched cards in a computing installation at one end of the spectrum to someone who know operations research, management science, and all of engineering rolled up into one. I do not know what a systems analyst is. He is somewhere in that spectrum, but I cannot say where.

QUESTION. Is there any information available on whether decision-making actually leads to statistically significant decisions?

ANSWER. I think you made a "no-no" there. Let us go back to the $5 and the $0–$100 coin toss. How can we measure in a one-shot decision what is statistically significant? This raises the issue of what view of probability we are taking. What we are saying is that the whole concept of statistical significance is pretty much irrelevant from the point of view of decision-making because we usually make decisions in one-shot situations. We cannot fire off a thousand Apollo rockets and see how many are going to succeed and how many are going to fail. We have to make a one-shot decision—do we go now, or do we not go now? And the question of statistical significance just does not come into it at all. I never find myself using those words. I find no use for them in making logical decisions.

QUESTION. Can you give some references for rating the qualitative effects of decisions?

ANSWER. You mean the so-called "intangible aspects"? There is a large amount of literature on the whole field. In general, what we say is that it is not a matter of tangibles and intangibles. If we take the Mexican example, people would say pollution is an intangible, or dependence on foreign supply is an intangible. But they are really tangibles. Why would you be willing to pay for things that are not tangibles? What we are saying is, let us take all the things that have value to you—positive and negative—and put values on them. In other words, if you would like to go out tonight and smell fresh air as opposed to smoggy air, let us talk about what that is worth to you. It is not worth $100 for you to do it for one night because you would go broke—that value would not be consistent with the other demands on you. But it is worth a penny, I will bet. Thus we begin to put dollars and cents values on what many people consider intangibles. Finally, we find ourselves making comparisons among values represented in dollars, not because dollars are in some sense the ultimate measure of everything, but because money is the Lagrange multiplier that our society has prepared for trading off one kind of thing against another kind of thing.

QUESTION. What are the axioms you must believe in order to reason this way?

ANSWER. Let me discuss just one of them. You must have transitive preferences, that is, you must reason such: if you like *A* better than *B* and *B* better than *C*, you must like *A* better than *C*. One of the points I was trying to make originally was that people often are not transitive. I might very well express to you intransitive preferences. But the question is, when you illustrate to me that I am being intransitive, do I like it or not? I do not like it, and the reason I do not like it is that someone can make a money pump out of me in that situation. I will switch *A* for *B* and *B* for *C*, and *C* back for *A*, all the while paying happily to make the transition—and he is just taking my money away, little by little. So the whole idea of intransitive preference is one of the things I do not want, and it is the cornerstone of what we do here, because the opposite of it is to be drained of your resources.

QUESTION. Could you comment further on the value of a life? Could you, for example, infer the value placed on life by observing the way corporations make decisions involving life?

ANSWER. Well, I have never done that. Of course there are studies all over the world on the value of a life. And unfortunately it varies greatly from one society to the next. But there was a comparative study I saw a few years ago indicating that at that time, for example, the value of a life was $100,000 in the United States, about $10,000 in Japan, and about $700 in South Vietnam.

There are many ways you can go about establishing a value for life. For example, you can examine the cash amounts awarded by juries for people killed in automobile accidents. The real issue is not what is the exact value of life, but rather are you being consistent in setting the value from one situation to another. The point is, life is precious, it is infinitely valuable. We are not talking about what you are willing to sell your life for—that is not the issue at all. The question is, what are you willing to buy it for. It is inconsistent to say a life is precious and, then, go out and not put the seat belt on when you get in your car. You are not being the same kind of person you would like to be at other times.

So what I like to do is pose a number for myself. (I cannot say what it should be for anyone else.) I want people to use this number as the value of my life when they make decisions that affect me. If I place it too high I will be running out of money very soon, because my car will weigh ten tons and will look like a tank. If I place it too low, I will not be able to venture outdoors.

You asked more specifically, could you determine from previous corporate decisions what value must have been placed on life. Well, first of all, I doubt that any corporation has ever established a number in the sense I am suggesting now. Perhaps they did it intuitively, but not explicitly. I would guess that if they did set such a value it would be $100,000 to $500,000 in the United States today.

QUESTION. Are the probabilities ever so hard to assign that you have to tell a client you cannot do an analysis for him?

ANSWER. No, I have never had that happen. Let me give you an example that arose in determining whether a new power nuclear reactor design should be introduced. The critical variable was the lifetime of the fuel cladding. The cladding was to be made from a material that had never been formulated before and, yet, the decision to go ahead on this new design would depend upon how this material performed.

The three people who were most knowledgeable in this company on the question of how long the material would

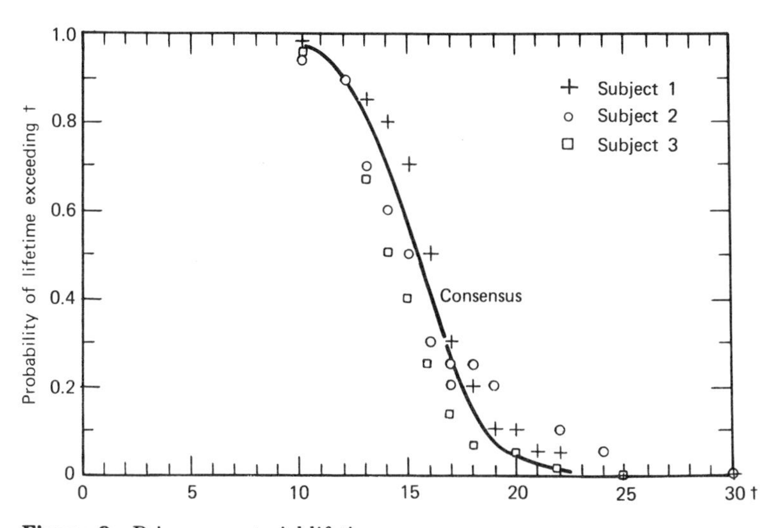

Figure 9 Priors on material lifetime.

last were assembled and told, "Look, we are going to have to come up with probability distributions on the life of this material, even though no one has ever built it before." They were interviewed separately, so they could not hear each other's answers. They gave the responses shown in Fig. 9. The first subject assessed about an 80 percent chance that the life would exceed 13. By an interview procedure, we developed the series of crosses. Then the next person (circles) was interviewed. He came up with the answers you see in Fig. 9, independently. Subject three's knowledge is represented by squares. Now when you look at these results, you find that they are remarkably similar. There are inconsistencies, however. For example, at one point in the questioning the subject represented by circles said that there was a 0.25 chance of exceeding a life of 17, and at another point, a 0.2 chance. There is a difference of 0.05 in probability. But when you think of how different these answers might have come out, I find their agreement remarkable.

Then the three men got back together again and started exchanging information. The last two subjects were relatively pessimistic about a short life compared to the first subject. He pointed out to them that certain things they were worried about were not really of concern because of experiments he had done recently. In other words, they exchanged information.

The same thing happened down at the far end. At the conclusion of the meeting they were willing to sketch the solid curve to represent the consensus of their opinion about how long this material would last. When this curve was shown to their boss, the manager of the whole operation, he for the first time felt that he was understanding what they were trying to communicate to him. He realized that they were not trying to be evasive, but were genuinely uncertain about what the life of this material would be, uncertain to the extent indicated by the consensus curve.

So, to answer your question, first I have never had the problem because you can always do it by means of a reference process. The basic idea is to say: would you rather win a million dollars, if the life of the material comes out more than 14; or win a million dollars, if a coin comes up "heads"? Then you can adjust the reference process, whether it is a coin coming up "heads" or a die coming up some number, until you are indifferent between the two processes.

But the point is that when a person is dealing with something every day of his life, when he is as expert at it as these subjects were, their answers are in much greater agreement than you would expect, if you asked them questions about the length of the Danube River, or other things they have not seen or thought about very much. What in the abstract seems as though it might be a problem, when you speculate about assigning probabilities, turns out to be not much of a problem, when you actually face a decision.

QUESTION. Can decision analysis tell whether it is worthwhile to do decision analysis?

ANSWER. You can do a "back-of-the-envelope" analysis, and then the question is how much would it be worth to do a more refined analysis. The decision to employ a decision analyst is, itself, a decision that can be analyzed like any other decision.

I have a rule of thumb that I have found helpful in my own life. I would like to spend at least one percent of the resources I am allocating on making sure that I am getting a good allocation of those resources. If I am going to buy a $2000 car, I want to spend at least $20 in making sure that I find the right car for me. Not making sure—in an absolute sense—but making sure in the sense that given the limits of my time and interest in the subject, I am doing the best I can.

In thinking about professional analyses, you should realize that decision analysts are in high demand. They command a premium professional salary. Consequently, decision analysis does not come cheap. However, we have not yet had a decision analysis that I am aware of where the decision maker did not feel very good about the insights he received relative to the money he spent.

QUESTION. Is it true that rather than removing intuition entirely from the process, you have simply removed it from the decision-making process and pushed it back to within the decision analysis?

ANSWER. Right. We are not eliminating judgment, or feelings, or opinions, or anything like that. Rather, we are quantifying them and putting them into a form where logic can operate upon them, rather than be buried in a man's mind where we cannot get access to them. This is very much a subjective and judgmental process in the sense that the probability assignments and all evaluations and preferences have to come from the decision maker. This is really just a matter of rendering unto Caesar the things that are Caesar's. What we say is, let the manager who makes the probability assignments and who has the preferences devote his time to making sure they represent his true feelings and his true attitudes. Let the logic be handled by the computer, which is eminently qualified to do such a job. It is really "divide and conquer". It sounds like a small thing, but the power of it is very great.

QUESTION. I do not see how a decision can be good unless the decision maker has good preferences.

ANSWER. Well, a decision is good if it is consistent with the decision maker's choices, information, and preferences. Looking on, I might think he is an idiot to make such a decision, but it is his prerogative. Our theory is amoral, in the sense that a person can go to Las Vegas, gamble, and lose money, but he says, "I have a ball there." "I value the experience very highly." Alright, given *his* values, he is making a statement with which I cannot disagree.

QUESTION. Would you comment on the balance between seeking new alternatives and analyzing the ones you have got?

ANSWER. Well, you cannot beat a new alternative. But how do you get a good new alternative? I find, for example, as an old engineer that there is nothing like doing an analysis of the existing design to see its weaknesses and to suggest improvements in that design. I think the same is true of a decision. We often find that we have two alternatives with the property that one is weak in one area, while the other is weak in a different area. Someone will suggest combining the two to create a new alternative with both good features. Often, this is feasible. These are new ideas that were not suggested originally by the individual who had the decision problem. There is no magical way of getting better alternatives by doing it this way, but it often turns out that creativity is a by-product of the process.

QUESTION. What makes you believe that you obtain better outcomes by decision analysis than you would by following intuition?

ANSWER. It is really an act of faith. Let us take the case of the man with the $5 payment for the $0–$100 coin toss. I say to him, "That is a good decision." "I have looked into your finances, and we agree it is a good decision." And he says, "Yes". He calls the coin and he misses, because after all he still has a probability of one-half of failing. Whereas some other person says, "I have looked into the Swami's eyes and I know I must call 'heads'." So he calls "heads" and wins. Which is the better decision-making procedure? It is really an act of faith that a logical procedure based on principles you believe in is better than another procedure. We can never prove that someone who appeals to astrology is acting in any way inferior to what we are proposing. It is up to you to decide whose advice you would seek.

QUESTION. Is it always possible to get better, more complete information and, hence, make a better decision?

ANSWER. Not always. For example, if a major hurricane bears down on Miami in the next hurricane season, where are you going to get more complete information? The decision will have to be made with the presently available information.

One of the persistent features of human nature is this quest for certainty. If anything too much money is spent on information-gathering, rather than too little. We keep pursuing this "Holy Grail" of certainty, instead of trying to find better alternatives or just making the decision and getting on to something else. I see the whole move toward data bases as symtomatic of this desire—this quest for certainty, hopeless as it is.

QUESTION. Is not a major function of an executive the ability to recognize when a decision has to be made?

ANSWER. Yes. For example, a president of a major company faces the decision of introducing a new product. He knows he has the decision, he is worried about it, and he does not know what to do. He has complex alternatives which are not easily evaluated, and he knows that intuition is not going to be much help to him. Therefore, he calls on a decision analyst to sort out the alternatives, get probabilities assigned, build models, present lotteries for his inspection, and so on.

QUESTION. The decision analyst will create the probabilities, will he not?

ANSWER. No, the probabilities do not come from the analyst but from the decision maker, his experts, and possibly external experts. I am not an expert on hurricanes; I am not an expert in medical problems; I am not an expert in the Mexican electrical system's rate of growth, or anything like that. God forbid, we should try to become experts in all the different things we work on. But what people in this profession are expert in is the modeling of the decision problem and the extraction of information from experts and preferences from decision makers in order to develop a better decision. It is a very careful separation of function.

QUESTION. If I were to adopt this approach and apply it to a variety of different decisions and if, after a while, I were to discover that this led to favorable outcomes less frequently than my old usual approach, then I would be forced to conclude that you were giving us a rather esoteric meaning of the word "good."

ANSWER. That is what you would be forced to conclude.

QUESTION. But does it not have to lead to more good outcomes, if it is to have practical value?

ANSWER. The question is: would you make the same decision if you faced the same situation again without knowing how it was going to turn out. I think it is a good decision to pay $5 for the $0–$100 coin toss. I would even be willing to purchase several of them. Suppose I keep losing. I would look at the coin, consider all kinds of hypotheses about people cheating me, two-headed coins, and sleight-of-hand, but suppose I am convinced there really is no "hanky-panky" going on. Well, then, I would not depart from this theory. I would say, "O.K., it is still a good decision, give me another one," even though I had lost five or six in a row—which is not an unlikely event. But I am going to stay with this theory until I find a better one, and I have not found one yet.

QUESTION. But should our goal not be to maximize the likelihood of good outcomes?

ANSWER. Of course. We all want joy. We all want good outcomes. Let that be stipulated right now. Everyone wants good rather than bad, more rather than less—the question is how do we get there. The only thing you can control is the decision and how you go about making that decision. That is the key. When you focus on that, I think you will want to do it the way we have discussed.

BIBLIOGRAPHY

Ginsberg, Allen S., *Decision Analysis in Clinical Patient Management with an Application to the Pleural Effusion Problem,* Ph.D. dissertation, Stanford University, July 1969.

Howard, Ronald A., "Decision Analysis: Applied Decision Theory," *Proceedings of the Fourth International Conference on Operational Research,* pp. 55–71, Wiley, 1966.

Howard, Ronald A., *Dynamic Probabilistic Systems,* Vol. 1, *Markov Models;* Vol. 2, *Semi-Markov and Decision Processes,* Wiley, 1971.

Howard, Ronald A., *Dynamic Programming and Markov Processes,* The M.I.T. Press, 1960.

Howard, Ronald A. (Ed.), "Special Issue on Decision Analysis," IEEE Transactions on *Systems Science and Cybernetics,* Vol. SSC-4, No. 3, September 1968.

Howard, R. A., J. E. Matheson, and D. W. North, "The Decision to Seed Hurricanes," in *Science,* (4040), **176,** 1191-1202, June 17, 1972.

Luce, R. Duncan and Howard Raiffa, *Games and Decisions,* Wiley, 1957.

Raiffa, Howard, *Decision Analysis: Introductory Lectures on Choices under Uncertainty,* Addison-Wesley, 1968.

Savage, L. J., *The Foundations of Statistics,* Wiley, 1954.

Schlaifer, Robert, *Analysis of Decisions Under Uncertainty,* McGraw-Hill, 1969.

Stanford Research Institute, Decision Analysis Group, *Decision Analysis of Nuclear Plants in Electrical System Expansion,* prepared for Comision Federal de Electricidad, Mexico City, SRI Project 6496, December 1968.

5

Divide and Conquer: How to Use Likelihood and Value Judgments in Decision Making

WARD EDWARDS

Professor of Psychology
Head, Engineering Psychology Laboratory
Associate Director, Highway Safety Research Institute
Institute of Science and Technology
The University of Michigan

Ward Edwards received his Ph.D. in experimental psychology. from Harvard University. Edwards is a Professor of Psychology and Head of the Engineering Psychology Laboratory at The University of Michigan, where he is conducting research on decision processes. His present work involves the application of utility theory to the evaluation of large social programs. He is a coeditor of the book, *Decision Making,* and a consulting editor for the *Journal of Experimental Psychology.* He is also the Associate Director of the Highway Safety Research Institute, Institute of Science and Technology, at The University of Michigan.

The Theme

The theme of decision analysis is "divide and conquer." Let us begin by examining what that means. Then we turn to instances—both pieces of research and applications—that apply that theme.

We start from the supposition that decision-making is and should be an inherently subjective human process. Many people suppose that once you say an intellectual task is subjective, that it is going to be done by people, that it depends on human judgment—you are through. Analysis ends there. I disagree. Inherently subjective processes are fully as amenable to scientific understanding and technological improvement as any other processes.

"Divide and conquer" summarizes a strategy for making such technological improvements.

The basic idea of "divide and conquer" consists of three steps. First, you take your cognitive, your intellectual task—whatever it may be—and you break it up into little pieces along natural lines of cleavage. What those lines of cleavage will be depends on the nature of the task. This is an analytic step. Second, you make separate judgments about each of the pieces into which you fragmented the task. These judgments are typically—always in the cases I work with—numerical. Third, you reaggregate the results of these judgments using an appropriate formal aggregation rule. This is a computational step.

The reasons why this is a good strategy for performing cognitive work are partially theoretical, partially experimental, and make more sense in context, hence I will not argue them now. The tasks to which this strategy applies fall into three broad classes. The first is the task of diagnosis: figuring out what is going on in our environment on the basis of inadequate evidence. (In the real world of decisions, information is almost always inadequate.) The second of the tasks is evaluation: assigning values to entities which may be outcomes, may be social or organizational programs, or may be anything. These values, then, are used in conjunction with the probabilities that are the output of the diagnostic process for the third of these tasks, which is decision-making, or choice.

Three different kinds of aggregation rules are relevant to these three different kinds of intellectual tasks. The aggregation rule that goes with diagnosis is Bayes's theorem of probability theory. The aggregation rule that goes with evaluation is nothing more complicated than the idea of weighted average. The aggregation rule that goes with decision-making is the idea of maximizing expected utility.

Diagnosis

Every serious decision maker faces the problem of diagnosis. The name, of course, suggests something that might be done in a doctor's office or a hospital, and indeed that is one context in which it is done. But it is also done by a business executive trying to figure out whether or not his firm should introduce a new product; by a military officer trying to figure out what it is that the enemy is doing preliminary to figuring out what he ought to do about it; by an intelligence analyst examining, in a political rather than a military context, much the same kind of question; by a government executive trying to figure out what the current state of some social problem is; and so on. It is, in short, a very widespread human activity.

Nearly every real-world, practical diagnostician I have ever met has the feeling that what he needs is more information, because he is uncertain. If he commands the resources and has the opportunity, as a result, he goes out and collects more. That, of course, is expensive. In addition there are many circumstances in which a decision maker just can not afford the cost in money or time. In some gamelike situations, collecting additional information is unwise because it may reveal your plans. Gathering new information often is not the solution to the problem of uncertainty—the problem of diagnosis.

The alternative is to make better use of the information already on hand. Let us use a military example. If we find military examples distasteful, we can easily translate this one into a business context. Compare the situation of a modern military commander with the situation of Alexander the Great to see how the technical resources of our society are brought to bear on the diagnostic problem. Alexander depended on the naked human eyeball as his basic information source. Twentieth-century commanders use that, but in addition use a wide variety of very fancy technical sensors. Alexander depended on men on horseback to carry the information to him. Today, commanders use electronic communications, traveling at the speed of light. Alexander depended on maps to record and display the information in a convenient form, so he could use it. Modern commanders also use maps, but they are often fancy electronic ones and are accompanied by other very elaborate displays. Alexander depended on men on horseback to carry away his decisions to those whose job it was to execute them. Modern commanders again use electronic communications.

But what about the central part of the task—the diagnostic part—the part that consists of looking at the information and figuring out what it means as a preliminary to figuring out what to do about it? How did Alexander the Great do it? He looked at the data, and maybe he scratched his head, and he made up his mind. How does the modern military commander do it? He looks at the data, maybe he scratches his head, and he makes up his mind. It sounds as if not much in the way of technological progress has occurred in respect to the central part of the task of diagnostic information processing since the days of Alexander the Great. That is true, but something can be done about it. To explain, we must retreat into abstractions for a while and come back to real problems shortly.

Probability

What is a probability? Of course, a probability is a number—a number between zero and one. Zero means impossible. One means certain. And if we admit that the coin is not going to land on its edge, and is not going to fall through a crack and not land at all, then we have to admit that the probability of "heads" plus the probability of "tails" is going to add up to one. Those properties, somewhat more formally stated, imply all of that beautiful body of formal ideas called probability theory. So we must have stated the question incorrectly when we asked what is a probability—and indeed we did. A better way of putting it would be: how can you recognize a probability when you meet it walking down the street? Or, in less fanciful language, what identification rules should one use to identify the abstract numbers called probabilities with events that one might observe?

One set of those rules is so familiar that it may never have occurred to us to question them. Suppose we wanted to estimate the probability of getting "heads" on a flip of this coin. We might flip it 100 times and observe 57 "heads." Probabilists of all persuasions would agree that the number 57 out of 100 has something to do with the probability of "heads." (To the best of my knowledge, no one since the psychologist E. G. Boring in 1941 has proposed that the number 57 out of 100 *is* the probability of "heads"; he did propose that in a remarkable article which treasurers of the curiosa of science might find entertaining.[1]) A set of identification rules for probabilities begins by noting that it was really an accident that we stopped flipping after 100 flips. We might have gone on flipping indefinitely. In generating a sequence of flips we generate a sequence of ratios of "heads" to total flips. Assume that the sequence of ratios approaches a limit as the number of flips increases without limit. We might then define the limit approached by that sequence of ratios as the probability of "heads." This is the frequentistic set of identification rules for probability and, as I say, they are so familiar that it probably never occurred to anyone to question them. They have some problems though.

One obvious problem concerns the conditions under which the observations are to be repeated. Suppose, for example, we were to flip the coin by bouncing it up and down in an open palm. Would one regard that as an appropriate way of estimating the probability of heads? Let us hope not. Going to the other extreme, suppose we used a

[1] E. G. Boring, "Statistical frequencies as dynamic equilibria," *Psychol. Bull.*, **48,** (1941), 279–301. Also in *History, Psychology and Science: Selected Papers,* R. I. Watson and D. T. Campbell, (Eds.), Wiley, 1963.

perfect coin tossing machine that always tosses the coin the same way each time, and we put it in the same orientation each time and used the same coin. Is that a good way of estimating the probability of "heads?" Presumably not—because if the machine tosses the coin the same way each time, we would get the same result each time. How, then, should you repeat observations? If you look at the fine print at the bottom of the appropriate page in your textbook of probability theory, you will find that observations should be repeated under "substantially similar conditions." What does that mean? It means conditions that are similar enough, but not too similar. Obviously, the task of determining when conditions are "similar enough" is rather subjective, and so the frequentistic set of identification rules is less objective than it seems. That is not to say that the determination is unguided by calculation. Every test of statistical significance is intended to answer the question whether conditions that differ are nevertheless similar enough so that they may be used to estimate the same probability. But tests of statistical significance are themselves rather subjective procedures.

Subjectivity of rules for repeating observations is not, however, the main objection to the frequentistic set of identification rules. The main objection is that most questions about which we are uncertain are not readily amenable to treatment in terms of relative frequency. So if we accept the frequentistic rules, we cannot use the formal equipment of probability theory to describe and manipulate such uncertainties.

Consider the following: I will be leaving Los Angeles. I will be leaving from an airport. Will it be Los Angeles International Airport? You can think of relevant relative frequencies. For example, of all the people who leave from all the airports in the Los Angeles region, what fraction of them leave from Los Angeles International? Now I will add another item of information. I flew here in my own plane. That information radically changes the probabilities, if you know that other airports are closer to Caltech than is Los Angeles International. But how do you think about that information in terms of relative frequency? I do not know.

A less restrictive set of identification rules is preferable—one which permits description of many more of our uncertainties by means of probabilities. Fortunately, such a set exists. Its modern resurgence is primarily due to L. J. Savage[2] who was, till his untimely death, a professor of statistics at Yale. In this view, a probability is an

[2] L. J. Savage, *The Foundations of Statistics,* Wiley, 1954.

opinion. Of course, if it is an opinion, it is some person's opinion. It describes that person more than it does the event. Such probabilities are called personal. I will remind you that probabilities are personal by speaking from time to time of "your" probability for some event or hypothesis. In so doing, I'm paying you a compliment, because not all opinions can be probabilities. Probabilities have to add up to one, and that is a very demanding requirement indeed—so demanding that no one person's opinions ever manage to fulfill it. So the "you" I speak of is a somewhat idealized you, the perfectly consistent you, you as you would like to be, rather than you as you are.

Bayes's Theorem

This is know as the Bayesian point of view. Bayes's theorem is a triviality—a direct and uncontroversial consequence of the fact that probabilities add up to one. But the conflict between Bayesian and classical approaches to statistics was until recently, and in some 'statisticians' opinions still is, very controversial indeed. Why?

Suppose you want to know the probability of a hypothesis, H_A, on the basis of some datum D. Bayes's theorem says:

$$P(H_A \mid D) = \frac{P(D \mid H_A)P(H_A)}{P(D)} \tag{1}$$

In words, to find the probability of H_A on the basis of D, take the prior probability that H_A had before D came along, $P(H_A)$, multiply that by the likelihood that D would be observed if H_A were true [here represented as $P(D \mid H_A)$], and divide by the normalizing constant $P(D)$.

A convenient working form of Bayes's theorem is obtained by writing Eq. 1 twice, once for H_A and once for H_B, and then dividing one equation by the other. The normalizing constant $P(D)$ cancels out, and the result is

$$\frac{P(H_A \mid D)}{P(H_B \mid D)} = \frac{P(D \mid H_A)}{P(D \mid H_B)} \frac{P(H_A)}{P(H_B)}$$

or, in more compact notation,

$$\Omega_1 = L \cdot \Omega_0 \tag{2}$$

Here Ω_0, the ratio of $P(H_A)$ to $P(H_B)$, is called the prior odds, Ω_1 is the posterior odds, and L is known in statistics as a likelihood ratio.

Why is Bayes's theorem important in spite of its obvious triviality? Because if probabilities are opinions, then Bayes's theorem is a formal rule that specifies how those opinions should be revised or modified in the light of new information. And the revision of opinion in the light of new information, otherwise called information processing, diagnosis, and the like, is one of the most important human intellectual activities.

Human Conservatism in Inference

That's all very well, but I am a psychologist, not a statistician or a mathematician. As a psychologist, whenever someone teaches me a rule that says how people should think, I am interested in comparing it with how people do think. I would like, therefore, to ask the reader to be a subject in an experiment relevant to that question. Imagine a bag full of poker chips. In my office I keep two bags of poker chips. One of them has 70 red and 30 blue chips. The other has 70 blue and 30 red. I flipped a fair coin to choose one, and here it is. The question is, what is the probability that this is the predominantly red bag. On the basis of the story so far, I hope you agree with me that it is 1:2. Of course probabilities are opinions and you are entitled to your own, but if yours differs from that number very substantially, we are just not communicating.

Now let us consider some evidence on the basis of which we can revise that probability. I will reach into the bag and mix them up, randomly sample one chip, replace it after recording its color, mix again, sample again, and do this 12 times. The result is 8 reds and 4 blues. On the basis of all the evidence, what is the probability that this is the predominantly red bag? Obviously, you could make a formal calculation. Do not do so, but instead use your intuition as best you can. You might find it a good idea to write your estimate in the margin of the page.[3]

The Reverend Thomas Bayes was an amateur probability theorist who had the immense good fortune to be in correspondence with the mathematician Laplace. I have never been able to find Bayes's theorem explicitly stated anywhere in Bayes's writings, but it is deducible from what he wrote. If Rev. Bayes had been a subject in his experiment, and if his intuition were well described by this theorem,

[3]When this demonstration was used during the lecture at Caltech, the answers ranged from about .55 to about .93; the modal value was between .70 and .80, and the overwhelming majority of answers were between .60 and .90.

which I doubt, the answer he would have given would have been .97.[4] That number is probably startling to you. The point of the illustration is that human beings are conservative processors of information, unable to extract from evidence as much certainty as the evidence justifies.

About 10 years of research has been done on the topic of human conservatism in inference. A typical early paper is Phillips and Edwards.[5] Recent critical reviews of this literature were published by Peterson and Beach[6] and Slovic and Lichtenstein.[7] A typical result from a 70–30 symmetric binomial experiment is shown in Fig. 1, based on data from Phillips and Edwards. Note that the spacing on the y axis of Fig. 1 is logarithmic. Such spacing and these independent and dependent variables make Bayes's theorem nicely linear. Human behavior is reasonably linear also. Not, however, the same line. Human beings are conservative processors of information. Much research[8] has shown that men are conservative information processors primarily because they misaggregate evidence. That is, they can judge quite well the diagnostic implications of a single datum. But they cannot put together a number of data properly. Their aggregation rule is wrong.

Typical laboratory experiments show that men waste 50 to 80 percent of the information available to them. Information is costly; the cost of such waste must be, and is, staggering.

[4] To calculate this, use Eq. 2. $\Omega_0 = 1$. For such two-hypothesis binomial cases,

$$L = \frac{P_A{}^r(1 - P_A)^{n-r}}{P_B{}^r(1 - P_B)^{n-r}}$$

where P_A and P_B are the probability of getting a red chip from bag A and bag B, respectively, r is the number of red chips in the sample, and n is the sample size. In this example, $P_A = (1 - P_B)$, and so

$$L = \left(\frac{P_A}{1 - P_A}\right)^{2r-n}$$

$2r - n$ is the difference between the number of red and the number of blue chips in the sample; sometimes symbolized $s - f$ (success minus failure). Thus $\Omega_1 = L = (.7/.3)^4 = 29.64$. Since $P = (\Omega/1 + \Omega)$, it follows that $P_1 = .9674$.

[5] L. D. Phillips, and W. Edwards, "Conservatism in a simple probability inference task," *J. Exp. Psychol.*, **72,** (1966), 346–354.

[6] C. R. Peterson, and L. R. Beach, "Man as an intuitive statistician," *Psychol. Bull.*, **68,** (1967), 29–46.

[7] P. Slovic, and S. Lichtenstein, "Comparison of Bayesian and regression approaches to the study of information processing in judgment," *Organizational Behavior and Human Performance*, **6,** (1971), 649–744.

[8] W. Edwards, "Conservatism in human information processing." In B. Kleinmuntz, (Ed.), *Formal Representation of Human Judgment*, Wiley, (1968), 17–52.

Division of Labor

But we know a formally optimal rule for information processing: Bayes's theorem. It wastes no information. Perhaps we can circumvent human conservatism by using Bayes's theorem to do our information processing.

Consider an example. She's beautiful, she's young, and it is your first date. You have, of course, many hypotheses about how the evening will go, but in order to keep the example simple, I will reduce their number to two, which I may as well label "yes" and "no." On the basis of your prior experience, you have some opinion about the probability of "yes", and if your prior experience is comparable to mine, that number is neither zero nor one. So this is a situation of genuine uncertainty, one in which you would like to process the information that comes along—and certainly not conservatively. Consider a datum. As you escort her away from her front door she takes your arm and snuggles up closely. Well, it is a datum—certainly not conclusive, one way or the other. On the other hand it is not utterly irrelevant. In short, it is just the kind of inconclusive datum that Bayes's theorem is

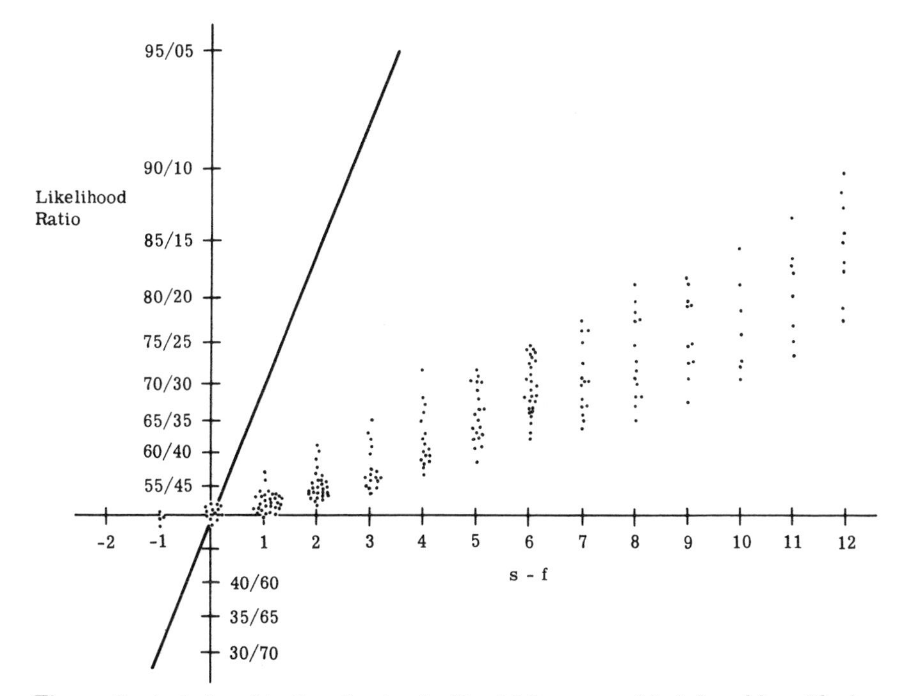

Figure 1 A single subject's estimates for P of 0.7, expressed in inferred logarithmic likelihood ratios as a function of the difference between the number of successes and the number of failures in the sample.

designed for. To calculate the needed likelihood ratio, we need the probability that she would take your arm if the outcome were going to be "yes", and the probability that she would do so if the outcome were going to be "no". We do not even need to know quite that much—we only need the ratio of those numbers. But neither those probabilities nor their ratio are readily available.

The difficulty, of course, is that in "bookbag" and "poker chip" examples we are dealing with what the mathematicians call a Bernoulli process—a formal model of a data generating process from which we can calculate those probabilities. But in these real world examples we have no such model.

The evidence indicates that human beings can judge very well the diagnostic meaning of a single datum. What they cannot do is put the pieces together. Bayes's theorem is a formal rule for putting the pieces together. And that suggests a division of labor between men and algorithms, in which men judge the diagnostic meaning of a single datum and Bayes's theorem does the aggregating. Divide and conquer.

Probabilistic Information Processing System

Figure 2 makes that idea a little more specific by means of a block diagram. Every information system begins with sources of information or sensors. These generate displays. Those displays are dealt with by people whose job it is to filter out irrelevancies and put what is left into some standard format and thus generate purified displays. The purified displays are looked at by the key men in the system, whose job it is to estimate a likelihood ratio for each datum and each pair of hypotheses of interest to the system. Those estimates are passed on to the Bayesian processor. If one is not in a hurry, a desk calculator will do just fine as a Bayesian processor. If one is in a hurry, he might prefer to use a computer. In any case, the Bayesian processor simply calculates Bayes's theorem. As a result it generates a display of the current probabilities of all of the hypotheses in the light of all of the evidence. That display is updated each time a new datum enters the system.

To remind you that the purpose of information processing is to be a prelude to decision making, Fig. 2 includes a decision maker. If you had a payoff matrix, you could have your decision made automatically by the principle of maximizing expected value. But generating a real-world payoff matrix is a difficult task, requiring a complex technology of its own.

The kind of system shown in Fig. 2 can be called a PIP, which stands for Probabilistic Information Processing system. Will PIP

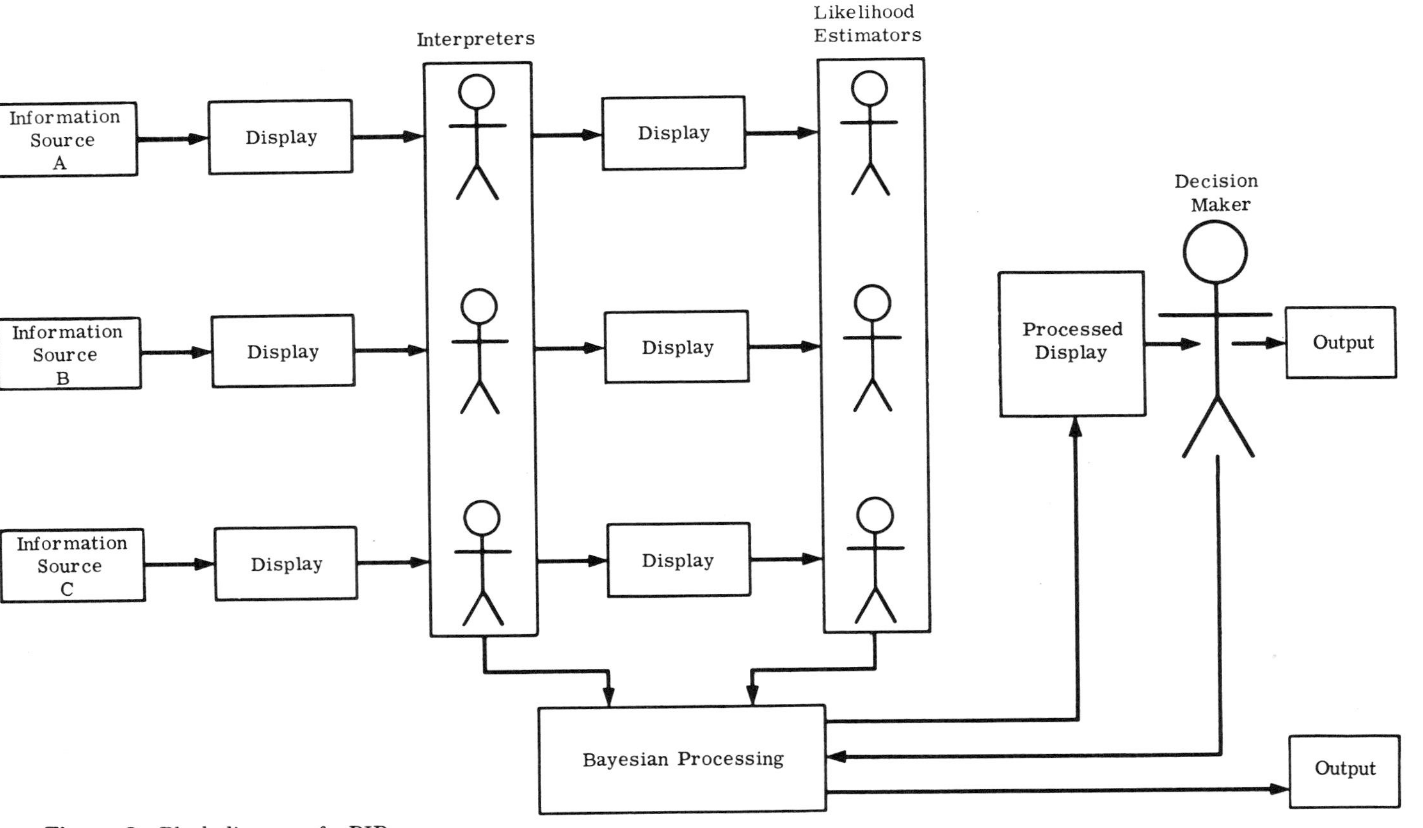

Figure 2 Block diagram of a PIP.

work? We wanted to find out in a relatively realistic context, so we designed, back in 1964, a world of 1975, connected to the world of 1964 by about 100 pages worth of history. 1975 had its own military technology, its own current political situation, and so on. The world of 1975 was designed to make six hypotheses reasonable. One of them was that Russia and China were about to attack North America. The second was that Russia was about to attack the United Federation of European States, which formed in 1969 after the break-up of NATO. The third was that Russia was about to attack the United Arab Republic. Another was that China was about to attack Japan. The fifth was that some other major conflict was about to break out. The sixth was that peace would continue to prevail. Data from three sources were relevant to these hypotheses. One was the Ballistic Missile Early Warning System (BMEWS), a large radar system capable of detecting missiles headed towards North America or Western Europe from the North Polar regions or regions beyond. A second was the intelligence system. A typical intelligence datum looked like this:

> Our agents have heard reports, credited to reliable sources, that a secret meeting was held within the last week somewhere in Outer Mongolia, involving Chinese Premier Liu and Russian Premier Balinin and their top aides. Also reported present were the top Russian and Chinese military officers. There is no evidence of any such meeting between these leaders in the past three years. Neither the Russian nor the Chinese press mentioned this meeting in any way, and, officially, Premier Balinin was reported away from Moscow for a "vacation" at his resort in the Ural mountains.

The third information source was a hypothetical photo reconnaisance satellite system. A typical datum from it looked like this:

> At 0800 this morning, 8 destroyer escorts sailed from Murmansk. They sailed due east and then, in the early afternoon, rendezvoused with a nuclear submarine. Since then, they have been conducting what appear to be antisubmarine warfare exercises.

How do we decide whether a diagnostic system is working effectively? If we do not know the truth, and in the real world we often cannot, you must compare one system with another. So we actually generated four systems: PIP, POP, PEP, and PUP. In PIP, of course, the subjects estimate likelihood ratios. We will not here describe the tasks of the subjects in the other three systems, except to say that, uniquely, in PIP, the aggregation of the evidence was done optimally in a computer by Bayes's theorem, whereas in all the other systems the subjects aggregated the evidence in their heads. POP, PEP, and PUP differed from one another only in the nature of the response mode used.

We started out with 75 subjects. We taught them for ten hours about the characteristics of the world, the current situation, and the data sources and the hypotheses. Then we gave them one of the toughest objective 2-hour exams I've ever seen, reduced their number to 36 of whom we later lost 2, so we ended up with 34 subjects. We broke them up into the four groups, taught each group how to operate its own information processing system and then presented each group with a total of 18 scenarios, each consisting of 60 data items. For each of these data items the subjects generated five estimates, and so we ended up with 34 subjects times 18 scenarios times 60 data items per scenario times 5 responses per data item, which produces 183,600 two-digit numbers.

Figure 3 shows the final odds for war after the 60th data item in each scenario. The first graph compares PIP and POP; the second compares PIP and PEP; the third compares PIP and PUP. All three graphs look about the same. The correlation between PIP and any other group is very high—.85 or higher. That is to say, the qualitative agreement between PIP and the other groups is very good. Although the qualitative agreement is very good, the quantitative agreement is not. PIP is much more impressed with a given amount of evidence than are the other groups. Or, to put it another way, the other groups are conservative relative to PIP.

Both axes in Fig. 3 are logarithmically spaced. Table 1, calculated from the regression lines of Fig. 3, shows what that means. Suppose that PIP were to see a scenario that would lead it to give odds of 99 to 1 in favor of some particular war. What odds would the other groups give? POP would give 4 to 1. PEP would give less than 2 to 1. PUP would give 4.6 to 1.

That is a rather dramatic difference in efficiency of information processing. Why? Because in PIP subjects judge only the diagnostic meaning of each separate datum, and the data are optimally aggregated by means of Bayes's theorem, whereas in the other groups the subjects must aggregate the data in their heads, and they do so conservatively. Divide and conquer. (For a fuller description of this experiment and of the PIP idea, see Edwards, Phillips, Hays, and Goodman.[9])

[9]W. Edwards, L. D. Phillips, W. L. Hays, and B. C. Goodman, "Probabilistic information processing systems: Design and Evaluation," *IEEE Trans. Syst. Sci. Cybernetics*, (1968), 248–265.

The story does not stop there because a lot of people became interested in applying these ideas. First came Dave Gustafson, whose Ph.D. thesis[10] was concerned with predictions about the length of stay in the hospital after a hernia operation. He compared PIP both with

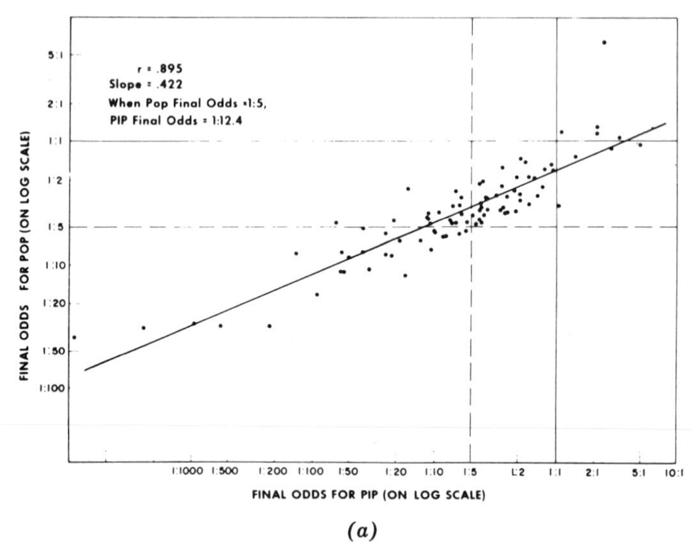

(a)

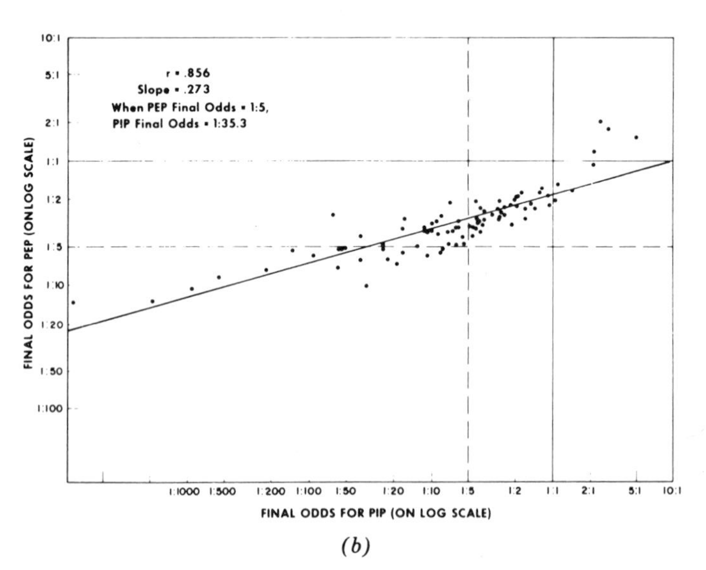

(b)

[10] D. H. Gustafson, *Comparison of Methodologies for Predicting and Explaining Hospital Length of Stay,* Ph.D. dissertation, University of Michigan, Ann Arbor, Michigan, 1965.

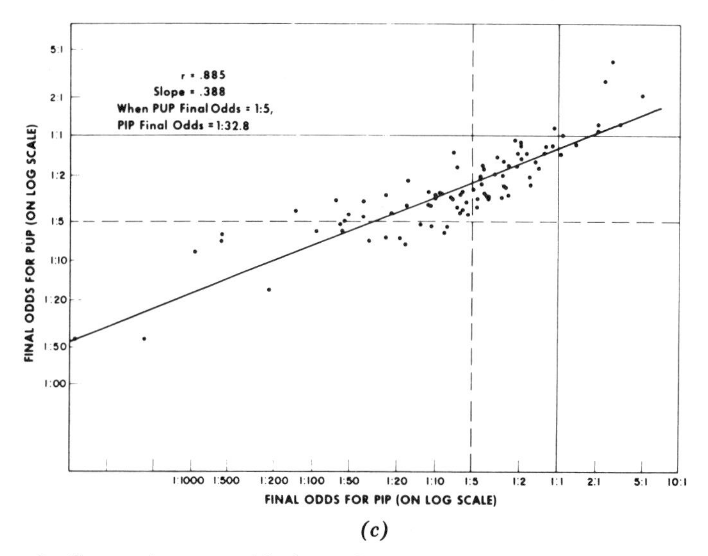

(c)

Figure 3 Geometric mean odds for each war at the end of each scenario. *(a)* POP versus PIP. *(b)* PEP versus PIP. *(c)* PUP versus PIP.

the doctors' direct unaided estimates, and with the best objective Bayesian and multiple regression procedures that could be devised. And PIP did better than either. Now he is at the University of Wisconsin engaged in a big project using PIP to diagnose thyroid diseases. (For more on PIP in medical contexts, see Gustafson, Edwards, Phillips and Slack,[11] and Lusted.[12])

At the University of Southern California, McEachern and Newman have used PIP to predict recidivism in juvenile delinquents.

A friend of mine in the intelligence community heard about all this, became interested, and decided to do some retrospective studies. He reprocessed the data of the Cuban missile crisis using the PIP technique and concluded that had PIP been used at the time, we would have had significantly earlier warning. Cheered on by that finding, he then proceeded to reprocess the data from the Chinese communist intervention after the Inchon Landing in North Korea, with even better results. (For an unclassified report of these studies, see Zlotnick.[13])

These studies processed data in cases in which the outcomes were known, and any kind of hindsight is suspect. The obvious thing to do is not to use hindsight, but instead to work with current situations.

[11] D. H. Gustafson, W. Edwards, L. D. Phillips, and W. V. Slack, "Subjective probabilities in medical diagnosis," *IEEE Trans. Man-Machine Systems,* **3,** (1969), 61–65.
[12] L. B. Lusted, *Introduction to Medical Decision Making,* Charles F. Thomas, 1968.
[13] J. Zlotnick, "A theorem for prediction," *Foreign Serv. J.,* **45,** (8), (1968), 20.

Table 1 Odds of Each War Hypothesis to the Peace Hypothesis

PIP	POP	PEP	PUP
99:1	4.0:1	1.9:1	4.6:1
19:1	2.0:1	1.2:1	2.4:1
1:1	1:1.7	1:1.9	1:1.3
1:5	1:3.4	1:2.9	1:2.4
1:19	1:6.0	1:4.2	1:4.0
1:99	1:11.9	1:6.6	1:7.7

Such work is in progress. (For an unclassified report, see Kelly and Peterson.[14])

Another obvious application is to the forecasting of weather. The weatherman reports various probabilities these days. They are generated by a very elaborate probabilistic computer model and may or may not be modified by local data accumulated since that computer model was last run. In any case, it seems very natural to apply the PIP technique to update such probabilities. Our first attempt to make such an application was abortive, but we plan later attempts.

Bayesian technology, a branch of decision analysis, has already been born and is growing. It is based upon numerical subjective judgments and on objective aggregation rules to put together the judgments. Thus, it illustrates the strategy of "divide and conquer." But that is not the only emerging technology based on that strategy. We will review another, concerned with evaluation.

Evaluation

In many different environments and contexts you may need to know the values of things. For example, operations researchers frequently would like a number to maximize. So would businessmen, for that matter. Operations researchers work very hard to find objective measures that they can maximize. The search for objective numbers to maximize is often preposterous. Every real world situation of even reasonable complexity embodies lots of different quantities one might like to maximize (or minimize). If we maximize one, we can not maximize others, because they are inconsistent with one another. We

[14] C. W. Kelly, and C. R. Peterson, *Probability estimates and probabilistic procedures in current-intelligence analysis: Report on Phase I,* June 1970–December 1970, Federal Systems Division, International Business Machines Corporation, FSC 71-5047, Gaithersburg, Maryland.

might like to fly an airplane, and do; and we might also like to drive a Ferrari—and would; but we can not afford to do both. There is a trade-off. Just that kind of trade-off among dimensions of value is inherent in every real world situation substantial enough to be worth thinking about. The hope to capture the essence of any really complicated decision situation in one single objective criterion is usually in vain and often silly.

A second kind of approach to the evaluation problem with which the author cannot sympathize is to treat real-world programs as though they were classical laboratory experiments. Suppose we are interested in evaluating the effectiveness of a head-start program intended to improve reading skills in underprivileged children. An experimental city gets the head-start program, and a matched (?) control city does not. Then we administer intelligence tests, knowledge tests, reading tests, or any kinds of tests we can think of to suitably chosen children in both cities. Then we do a T-test, and if there's a significant difference in the right direction, head-start is great, and if there isn't, down with it.

Such an evaluation procedure is absurd for a variety of reasons. The most obvious one is that there is no such thing as a control city that is similar to the experimental city; if there were, that city too would have gotten a head-start program. At least, its politicians must be less competent.

Utility

How can major social events, whether they be the results of experimental programs or experiments in space or indices of water pollution, be evaluated? We must start by recognizing that values are inherently subjective. The name given to subjective value, back in the Seventeenth Century, was utility. So the problem is one of the practical measurement of utilities. Economists tell us the following things about utility. First, it should be maximized; more specifically, the decision-maker should maximize his own utility. Utilities are not inter-personally comparable. They are measured, at least in decision-making under uncertainty, up to a linear transformation, or on an interval scale—they have no true zero point. And utilities are independent of the acts being considered. Those are all classicial properties of utility. The essence of what we have to say about utility is the negation of all of them. Utility as we think of it, should certainly be maximized, but there is no such thing as a decision-maker. The utilities we talk about are social, not individual and, therefore, if not interpersonally comparable, they are at least interpersonally generated. They are measured on a ratio scale and, therefore, do have a true zero point of a sort (not a very good sort) and they depend very much on the acts being considered.

How can we measure utilities? First, we must answer two questions. Utility for what purpose? And, whose utilities?

Asking questions about utility for what purpose points out the fact that what utility we are interested in depends on what decisions we have to make. For example, one context in which the author has worked on these ideas is the selection of experiments for inclusion in manned space flights. It turns out that three different utilities for each experiment are needed in that context. They are relevant to the selection of the experiments, their scheduling, and, when something goes wrong, their rescheduling. For example, the weight of the apparatus is an obviously important issue in the selection of experiments to take along. But it is completely irrelevant to the scheduling or the rescheduling, given that selection decisions have been made. So we must determine what decision is to be made before we can start talking about what utilities are relevant.

Next question: Whose utilities? Obviously, the decision maker's. But there never is a decision maker for any really serious public question. Or to put it another way, there are too many. The decision maker in principle is some very high official of his company or his government. So he is guaranteed to be too busy to be fully informed about the decision he has to make. He therefore relies on his staff. They must assemble the alternatives and the information bearing on values of each for him to use. His staff attempts to think as he would think. They do not use their own utilities, they use their best guess about his. He does not even use his own utilities. He is not in the game for himself—he is in it for the organization that he serves. In short, a sort of organizational utility gets maximized.

Suppose we know what decision we are considering and whose utilities or what organization's utilities should be maximized. Next we must identify the entities we are going to evaluate. Every real-world decision is a decision under uncertainty, so in principle those entities would be outcomes, dependent both on the action and on what state-of-the-world turned out to be true. In practice, we will frequently ignore this and simply evaluate programs or acts without attending to the fact that they really are decisions under uncertainty. We can not defend that practice, but we will not attack it either. It seems inevitable.

Next we identify the dimensions of value. Values really are arranged in a sort of hierarchical tree structure. Up at the top is "the good life," or something of the sort, and way down near the bottom is the fact that we prefer steak to lobster for dinner because we might not sleep so well, if we ate lobster. How far down that tree structure we want to go in defining the dimensions of value is only partly arbitrary. The general principle is: do not work with too detailed and specific a set of dimensions of value or we will have too many dimensions and then none of them will be very important.

Then, with respect to each dimension, judge the importance of that dimension to the aggregate, maintaining ratio properties for these importance judgments. People can make such judgments. In fact, they enjoy it. Next, normalize these importance judgments, which preserves their ratio properties, but makes them add up to a given number (which we can choose as 100 for convenience). Then, measure the location of each entity on each dimension. After measuring the location of each entity on each dimension, we may need to rescale these locations because many of these dimensions are objective; and the number we get depends on the unit of measurement. For example, we get a different number for the weight of the apparatus for a space experiment depending on whether we are measuring in pounds or in grams. And, yet, we do not wish that difference to enter into your evaluation process. So we have to rescale. There are techniques for that. Having rescaled, we can then do the simple arithmetic which consists of multiplying the number which represents the location of the entity on the dimension by the importance of the dimension and adding these products to get the aggregate utility. (For a fuller exposition of this approach to utility measurement, see Edwards.[15] For a related but different approach, see Raiffa[16]).

An Evaluation Example

We tried this procedure a few weeks ago with a group of planning officers in several police departments serving a large city. As an example they chose to work on a manning decision: several alternative manpower allocations among four police stations. As it happens, one officer present was on the planning staff of the local chief of police and another was on the planning staff of the local sheriff. The chief and the sheriff are rather different people and it was felt that they would evaluate these manning plans differently. So we invited each planning staff member to play the role of his boss. We then measured

[15] W. Edwards, "Social Utilities," Proceedings of a Symposium, "Decision and Risk Analysis—Powerful New Tools for Management," U.S. Naval Academy, Annapolis, in press.

[16] H. Raiffa, *Preferences for Multi-Attributed Alternatives,* The RAND Corporation, RM-5868-DOT/RC, April 1969.

the utility of each manning plan twice, once for the chief and once for the sheriff. The sheriff's importances all came out, before normalization, very much higher than the chief's. Of course, after normalizing, that difference did not matter a bit; it was only a stylistic difference. And, yet, that is what had led the policemen present to expect that the chief and the sheriff would evaluate the actions being considered very differently. After normalization, it turned out that there was only one inversion of importance on the dimensions that had been identified. The sheriff thought that pleasing the community—otherwise known as public relations—was a little more important than the chief did. Only a little bit, though. By the time we went through the whole procedure and came out with evaluation of the four manning plans we were considering, it turned out that the chief and the sheriff ranked these four plans identically by this procedure, and ranked them differently from what their aides' intuitive judgments would have led them to expect. So where we expected to get interpersonal disagreement, we got interpersonal agreement.

These procedures can produce interpersonal disagreement, of course. When they do, what they offer is essentially a set of rules for debate. They help pinpoint where the disagreement comes, why it comes, and enables us to talk numerically about its size. And we can discover by sensitivity analysis just how important the disagreement is. In any case what we have done is to take the problem of values apart by identifying separate value dimensions, to make judgments separately on each value dimension and judgments separately about the importance of each value dimension, and then to re-aggregate by means of the simple notion of a weighted average. Again, "divide and conquer."

Conclusion

We do not have examples of the successful application of this evaluation technology yet, though similar ideas are familiar to operations researchers. But we believe that it contains the seeds of another technology. And the technology is not very different from the PIP technology.

So, "divide and conquer." Fragment the task along its natural lines of cleavage. Make judgments about each piece separately and, then, use a computational algorithm to put those judgments together in order to come up with the desired output.

Question Period

QUESTION. It seems to me that you have amply demonstrated that the probability you calculate from PIP will pretty thoroughly fly in the face of intuition, and this would seem a good way to give a manager ulcers. Do you have suggestions on how to get a manager to accept them?

ANSWER. The best way is to have him come to you to ask. It is a lot easier if he asks you, than if you have to sell him. There are various ways in which you can attempt to make the case. One is that many people can recognize, at least in others, the phenomenon of human conservatism in information processing. I suspect that most of you find that idea intuitively plausible and can look around at others and see many instances in which they failed to exploit all the evidence available. This suggests that maybe formal processing will help. Another way you can go at it is simply to make the argument that after all, Bayes's theorem is a consistency rule. It specifies formal interrelations among different kinds of opinions. You can therefore exhibit, if you ask him to make appropriate numerical judgments, the inconsistencies that will appear. Then, if he is willing to believe that consistency is desirable, that invites him to start worrying. At least some of his judgments are wrong; maybe they all are. Of course in the last analysis the proof of the pudding is in the eating and, so, you invite him to try it out, without having his whole operation depend on it. He accumulates experience.

Actually, you have a preliminary selling job to do before you can even start to sell PIP, and that is to sell the idea that uncertainty is quantifiable. Indeed, many people believe (although I happen not to) that this is a more important thing to sell than the aggregation rule itself. Whether that is true I do not know—both must be sold, of course.

QUESTION. Have you made any attempt to correlate decisions made with the PIP process with decisions reached by an individual who has been arbitrarily defined as a successful top-level executive?

ANSWER. Not really. I have given talks about conservatism to audiences of mathematical statisticians and used with them the same example I have used with you of eight reds and four blues, and I always get identical results. That does not really quite speak to your point. It is rather difficult to arrange a comparison of PIP with a business executive. Our best successes have come with intelligence analysts, who perhaps face more uncertainty than do many business executives. There the general story I'm talking about shows up very clearly.

QUESTION. When you talk about "divide and conquer," have you had any experience in trying to find out the relationship between the division of resolution and the success of PIP? In other words, you can "over-subdivide" something, and all of a sudden PIP becomes meaningless.

ANSWER. Quite. The real problem latent behind what you have said is the problem of independence. What you really want to do is to fragment things in such a way that they meet the test of a condition called "conditional independence." I will not go into the formal technicalities of that; but, in general, data come to you in natural units anyhow, in most information processing situations with which you deal. Those represent the finest subdivision that you would want to consider. Frequently you will use coarser ones than that in order to get conditionally independent clusters. It turns out that people can judge very well just what items of evidence should be treated together in order to get a conditionally independent cluster. There is research on that by myself, by Manley at Ohio State, by Dave Gustafson, and it all indicates that people do this quite well. That is essentially a judgmental matter, but people are pretty good at making that kind of judgment.

QUESTION. Could you clarify the results of the comparison between PIP, PUP, PEP, and POP? It seems to me you have to add psychological conservatism to the results before you can say that PIP works better. Did PIP come closer to the prior probabilities?

ANSWER. You do not want to come close to a prior probability. You want to get as far away from it as the evidence justifies. A prior probability is something to be revised as evidence comes along. I think what you mean is that there was no definition of what the right answer was—what hypothesis was true. That is correct. A number of experiments have examined the PIP procedure in situations in which the data generating process is simple enough so that you can describe it mathematically, and at the same time is somewhat realistic. One such experiment was concerned with heights of men and women. These are fairly well describable as two normal distributions with the same variance, but different means. You can say, we will choose either men or women and we will sample successive individuals and we will tell you just their height. The question is, what are the odds that these are the men rather than women that we have sampled. You do this kind of an experiment and with the PIP procedure you find they come out very well. Right up there where the odds ought to be. And with any kind of direct estimation procedure—estimation of the posterior quantity—they are very conservative. If you do enough experiments of that general character, where you do have a model of the data generating process and therefore can calculate the right answer, you come to have a feeling that even in situations where you do not have a model of the data generating process and therefore can't calculate the right answer, you are going to get the same sort of thing you have found in these others. What you find in all the others that I know of, is that PIP is somewhat up there in the right ballpark and anything else, where man must aggregate in his head, is quite conservative.

QUESTION. Where in society would you like to see these techniques applied that are not being applied now? Where do you feel would be the greatest gain in utilizing this approach?

ANSWER. I suppose that in a certain sense my own activities provide the best answer to that question. I am certainly trying to get this technology used in the kind of information processing that underlies governmental policy making at as high a level as I can manage to reach. That is one way of answering your question. Another way is, if you look around for the kind of activity where conservative information processing is done that in the aggregate has the most effect on our daily lives, it is obviously going to be business. So this kind of technology works all over business; I do not see why it should not be used there, and I am promoting it where I can.

BIBLIOGRAPHY

Edwards, W., "Why Machines Should Decide," *Innovation,* **5,** (1969), 34–41.

Edwards, W., and A. Tversky, (Eds.), *Decision Making: Selected Readings,* Harmondsworth, 1967; Penguin Books, 1967.

Raiffa, H., *Decision Analysis, Introductory Lectures on Choices under Uncertainty,* Addison-Wesley, 1968.

6

Analysis Techniques for Operations Research

PHILIP M. MORSE
Professor Emeritus of Physics
Department of Physics
Massachusetts Institute of Technology

Philip M. Morse is the President of the American Physical Society, Director Emeritus of the Operations Research Center of M.I.T., and a member of the National Academy of Sciences. He was the first president of the Operations Research Society of America and in 1946 received the Presidential Medal for Merit of the United States. Morse has written extensively on the subjects of physics and operations research. He has authored *Queues, Inventories and Maintenance* and *Library Effectiveness: A Systems Approach.* He has coauthored *Quantum Mechanics* with Kimball, *Methods of Theoretical Physics* with Feshbach, and *Theoretical Acoustics* with Ingard. He is a coeditor of *Operations Research for Public Systems.*

We often hear the criticism that "we can send a man to the moon, but we can't run a city" or "we can build an efficient TV set, but we can't run a railroad." Implicit in these catch phrases is the question: "if the methods of physical science have been so successful in designing equipment, why can they not be used in handling some of our social problems?" The subject of this chapter is essentially a discussion of this question.

The Methods of Physical Science

First, a description of the methods of physical science. Simply, it is a matter of observing the phenomena and, then—instead of blindly amassing data—of forming a theory, a quantitative model of the essence of the phenomenon. This theory must then be tested by a series of controlled experiments suggested by the theory, which will either disprove the theory or will verify it and, then, will evaluate the parameters assumed in the model. Eventually, if the theory has been proved, we can use it to predict. And this is the practical advantage of the method, for if we can predict how the phenomenon will behave in various situations, then we can design equipment which will behave as desired.

Another aspect of the development of physical science is that it starts small. We start with a little piece of some aspect of nature and try to understand that piece before going on to larger aspects. One does not begin designing a moon rocket before understanding the elements of dynamics, electronics, and hypergolics needed to make the system perform.

Already we can see some answers to our introductory questions. The methods of physical science are successful only when the phenomenon studied has enough continuity and regularity so that experiments can find quantitative results, which can be repeated as desired. We do not need perfect repeatability; the methods of probability can ferret out regularity behind random fluctuations, if there is any regularity. But the phenomena cannot be one-shot affairs or situations where the environment cannot be duplicated, which, of course, rules out many of the broader social and political problems.

But there are social—or semisocial—activities that are repetitive. Many of these are conditioned more or less by physical boundaries or by the limitations of equipment, and many of them are directed at goals which can be measured quantitatively. One thinks of production lines in a factory, of air traffic—indeed of any kind of transport—of various kinds of operations of war, such as bombing. These repetitive actions of teams of men and equipment, directed toward specific ends, are now called *operations*. They abound in our society: fire fighting, rubbish disposal, hospital operation, marketing; we encounter them or take part in them every day. They can be studied by the methods of physical science; they can be observed, quantitative theories can be devised about them, predictions can be made about them, and, to some extent, they can be designed to produce the required result most efficiently.

Mathematical Models

Of course the degree of variability in these phenomena is usually much greater than with simpler physical phenomena. But this just means that the theory of probability plays a much greater role in the mathematical models of their operation than in most physical theories and that adequate prediction must include the range of variability of the outcome, as well as the average result.

In this new application of the method of physical science—which is called *operations research*—we have had to start small. We have to understand the details of a complex operation before we can start to predict its overall behavior. Let us begin with a few of the simpler examples and, later, examine how they may be put together. As with physical science, we find that mathematical models devised for one simple operation can then be applied to many other cases.

Queuing Models

For example, there is the operation of queuing up for service (see Fig. 1). Units—people, planes, radio sets—arrive at some point in space or time where they need service: the people at a checkout counter of a supermarket, the planes at an airport, the radio set with a burned-out tube. The arrivals are irregular, but with a certain average rate—λ arrivals per hour, day, or year. And they are serviced at an average rate—μ, though there is usually much variability in the service times as well.

The first thing one observes is that a model using just averages produces worthless results. With such a model, if the arrival rate λ is less than the service rate μ, no problem is expected—the flow would be unimpeded and the service facility would be busy a fraction of the time, equal to (λ/μ). Yet, we know that in most of these operations a queue is formed of units waiting to be serviced, even though the average service rate is capable of keeping ahead of the average arrival rate. The reason, of course, is the variability of arrivals and of service times. There will be times when the arrivals come closer together than the average and there will be times when the service takes longer than usual. The person ahead of you at the checkout counter may have $20 worth of groceries and you may have a 25¢ item, but you have to wait in the queue. Here is an example where the mathematical model must include the degree of variability of the operation before it can predict the sorts of delays that can occur.

For example, if arrivals are purely random (so-called Poisson arrivals) and if the service times are equally randomly distributed, the average length of the queue is $\lambda^2/\mu(\mu - \lambda)$ and the mean delay in queue is $\lambda/\mu(\mu - \lambda)$. On the other hand, if arrivals are random but each service takes exactly the same time, $1/\mu$, then the queue size

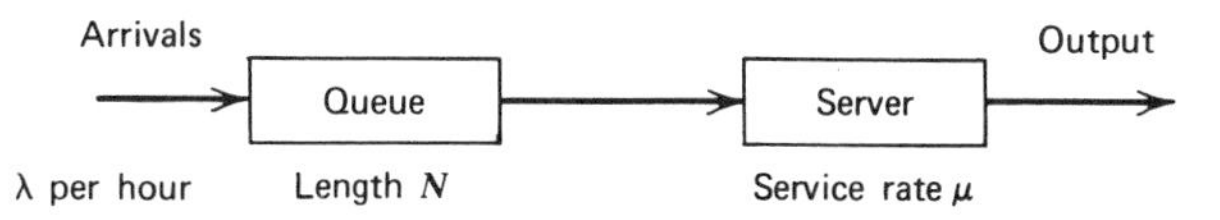

Figure 1 A queuing model. For random arrivals and random service times the average queue length is $\lambda^2/\mu(\mu - \lambda)$. The average delay in queue is $\lambda/\mu(\mu - \lambda)$.

(and the wait in queue) is cut in half. And if the arriving units come in bunches, with the average size of the "bunch" being M units, then the mean queue size is $M\lambda^2/\mu(\mu - \lambda)$, and so on. The theory also predicts the variability of queue size and delay times.

In every one of these formulas the factor $(\mu - \lambda)$ occurs in the denominator. The ratio (λ/μ), which equals the mean fraction of time the service facility is busy, is called the *service utilization factor.* As it approaches unity, all of these formulas increase without limit. Of course if arrival rate λ is greater than service rate μ, the queue does increase without limit and no steady-state behavior is ever reached. But even if (λ/μ) is less than unity, the delays can become too large.

The sensitivity of queue size to (λ/μ) when this nears unity is a factor that managers of operations find difficult to allow for. Discovering that the service facility is only working three-quarters of the time, they first try to reduce idle time, without realizing that increasing (λ/μ) from 3/4 to 7/8 increases the size of the queue and the waiting time by factors of nearly 3. During the early part of World War II, the port of New York was judged to be operating at about three-quarters of its maximum capacity in loading and unloading ships. In the interests of efficiency and to reduce loss to U-boats along our coast, it was decided to service in New York all those freighters which ordinarily went on to Philadelphia and Baltimore. This addition changed the utilization factor (λ/μ) from 3/4 to about 9/10, and (in accord with the formulas) increased the delay in port by a factor of 3 to 4. The members of the shipping board, who were thinking in terms of average rates, at first could not believe that a small change of 15 percent in utilization could cause a change of over 200 percent in delay and were certain that sabotage was the reason. But they were finally convinced that the change was to be expected, canceled their previous orders, and allowed shipping to go on to the more southern ports. Within a few months, delays at New York had returned to "normal" and, of course, coastal sinkings had again increased.

This example illustrates several of the problems inherent in applying operations research. One is the problem of "suboptimization." In a queuing system, if you reduce idle time—that is, reduce $1-(\lambda/\mu)$—you automatically increase queue size and delay in queue. If the manager of the service facility does not care about these delays, he can make the change. But if the facility is a part of a larger system, the delays may *reduce* the efficiency of some other part of the system. If the facility is part of a production line, for example, it may reduce the overall throughput. Or if the facility is an airport, the "queue" of planes stacked above it, waiting to land, may become dangerously large. And so, in a complex system, one part of the system may have to run at reduced efficiency in order that the system as a whole operate more efficiently. The service facility manager may have to leave his crew idle a quarter of the time for the whole operation to behave efficiently.

Another problem is that of deciding on the criteria for measuring the effectiveness of the system. If utilization of the service facility is to be the criterion, then delays in queue will have to increase. But if optimal throughput is the criterion, then the service facility will have to operate at reduced efficiency. This question of the appropriate measure of effectiveness is a basic one in all applications of these studies. In most cases the decision must be made by the general manager of the operation, not by the subordinate manager or by the operations analysts, and often the decision must be based on nonquantitative grounds. Should the success of a strategic bombing operation be measured by the total number of bombs dropped, by the amount of destruction produced, or by the political and psychological effects of the bombing? These are questions which must be decided at the highest level, and the decisions will strongly affect the way the operation is planned and carried out.

Such problems make it necessary for the team of operations analysts to have direct access to top management, so that the manager can come to understand all the implications of the analysts' predictions, the analysts can come to understand the manager's basic goals, and suboptimization can be avoided as often as possible.

An important part of the job of the analyst engaged in applying operations research is educational. He must give the manager a true understanding of the implications of his analysis. And he also must recognize that occasionally his recommendations will be overruled for political or other nonquantifiable reasons. In such questions of broad strategic goals, operations research plays an important, but only subsidiary, role. In fact, operations analysts may quite legitimately differ in their choice of the appropriate measure of effectiveness which is to be optimized. The final decision must be the responsibility of top management, once the quantitative implications of the choice are clear.

Markov Models

Queuing theory has now become a burgeoning branch of applied mathematics. Applications are found all over: in the fields of communications, of maintenance, of transport of all kinds, and of inventory control (where an "arrival" is the removal of an item from inventory and the "service period" is the interval before another item from the factory replaces this item in the inventory). In each case, when one determines the statistics of the arrival and service processes, the formulas predict the characteristic behavior and its degree of variability, so a balance can be struck between policy requirements for the system as a whole.

Queuing theory is closely related to the theory of Markov processes, which is a sort of probabilistic dynamics. Here the "driving forces" are the transition probabilities, the chance M_{mn} that, if the system is initially in state m, it will be in state n a unit time later. The analogy with classical dynamics is close. If the matrix of the M's is independent of time, the system may continue to oscillate between states or it may settle down to a steady-state behavior. Of course, the "steady state" is just in terms of the probability distribution between states; the system will be jumping from one state to another in accord with the M's. Nevertheless, one can talk of a "transient behavior" when the "forces" (the M's) are suddenly changed. Many queuing processes can be analyzed in terms of Markov theory.

To move to another field of application, the Markov model has been used to represent the circulation of books in a library. In general, the popularity of a book diminishes with time; the average yearly circulation of a class of books decreases more or less exponentially to some minimal value as time goes on. But average values do not tell one enough about the circulation history of an individual book. Its circulation may suddenly increase—because the subject has become more popular or, in a university library, a new course has been introduced—after which the circulation will again decrease. If one wishes to predict the distribution in circulation of a class of books, one must take these exceptions into account, and the Markov model will accomplish this. For example, if space requirements force the storage of a group of low-circulation books in some less accessible space, the model can estimate the number of library users who will be inconvenienced when they wish to borrow one of them. Many other systems, of greater social importance, can be represented by Markov models.

Other Models

In most operational situations there are a large number of parameters, some of which must be measured and some of which must be fixed by policy considerations. This brings in the large field of combinatorial mathematics. For example, a petroleum company can buy different sorts of crude oils and can adjust their refineries to produce different amounts of products, such as gasoline, fuel oil, and lubricants. The sales department makes estimates as to the amounts of these products they can sell during the coming year. The company then has the task of deciding the amounts of crudes which it should buy and the settings of the refineries so as to produce the required products at the least cost.

This is the problem of constrained minimum—or maximum, as the case may be. The objective function is a combination of the adjustable variables. In the oil case it is the sum of the costs of the crudes and of their processing which is to be minimized—subject to various constraints, such as the estimated requirements and the capacities of the refineries. The method is called *mathematical programming.* If the objective function is a linear combination of the variables, it is called *linear programming.* The method can be applied to many problems of allocation—of funds, resources, or manpower. The process usually involves many variables and specific solutions can usually be obtained only by the use of a large computer. The theory is concerned with programming the computer so that solutions can be obtained with the least amount of trial and error. When time is also involved, as in the complex problems of the scheduling of airline crews and planes, the analysis is still more complex. Such problems may tax even the largest computers. Some of the applications of this method to public systems are in the allocation of police patrols, the scheduling of fire-fighter personnel, and the scheduling of hospital operating rooms.

When the constraints are not precisely determined, the problem becomes still more complex. In many of these cases it is best to utilize Markov theory and work out the solution by the method called *dynamic programming.* Some of the related problems have, as yet, no method which is guaranteed to produce the answer in a finite number of tries. The "traveling salesman problem," for example, is of this type. This is the problem of laying out an itinerary between 20 to 50 cities, spread over the country, which minimizes the travel cost. Specific cases have been solved, but no guaranteed general method of solution has yet been developed.

Simulation

These are some of the analytic techniques which have been developed to represent—and thus to predict—various elements of an operational system. When it comes to combining them so as to predict the behavior of the whole system, the calculations are usually so complex that they can only be carried out on a computer, for each assumed set of values of the constraints and for each alternative set of operating policies one may wish to decide on. Thus, if the behavior of the various elements of the system is well understood, one can *simulate* the behavior of the system on the computer and carry out surrogate experiments which could not be carried out on the real system for reasons of cost, time, or morale. Computer simulation of complex operational systems is a big business now. When carefully carried out, with all the subsystems and their interrelations well understood, it can be of great value to the administrator. There are correspondingly great dangers, of course, if one does not understand the behavior of some of the components or if one uses overoptimistic forecasts of the performance of some new equipment, for example. The computer people have a word for it, GIGO—garbage in, garbage out.

U.S. Army Inventory

A few examples may show how simulation can be used. The spare parts inventory of the U.S. Army is a very large and complex system with thousands of different items flowing from factory to major depot, to theater depots, and eventually to the supply sergeant, who hands out the items as needed. Its total value is several billion dollars. Each node in the network operates according to a set of rules which specify at what inventory level more of an item should be ordered from an upper-level or equal-level depot. The managers of this system suspected that the total inventory carried was too large and asked the M.I.T. Operations Research Center to look into the question.[1] The choice of appropriate level, of course, is determined by a balance between cost and the probability that a supply sergeant will be out of an item when it is needed. Because of the variability of demand one can never have an inventory large enough so that he *never* is out of an item. The decision as to what chance of outage will be allowed is a policy question which must be decided by top management (after looking at the allocated budget). Once this is decided for each item, the queuing equations can be determined; and, at each node, the appropriate set of ordering rules and the mean inventory level can be established, *if* one knows the distribution-in-time of the demands from the lower levels. The operating rules then in force assumed that demands at each node arrived at random, in a so-called Poisson distribution.

The M.I.T. Group, H. P. Galliher and two students, first studied the statistics of the system the Army was using and found some peculiari-

[1] H. P. Galliher, P. M. Morse, and Mary Simond, "Dynamics of the Classes of Continuous-Review Inventory Systems," *OR Journal,* **7,** No. 3, (June 1959), 362-384.

ties indicating that the rules were not being followed very well at the lowest level. Apparently the supply sergeants had never been told about the Poisson distribution. For example, with an item—a carburetor, perhaps—having an average demand of 6 items a week, there is a very good chance that the demand will be 3 one week and 9 another week. When the 9-demand week came along the typical sergeant would forget about the 3-demand week and order as though the mean demand had increased by 50 percent.

It did not seem practical to educate all supply sergeants in the intricacies of the Poisson distribution, so it was necessary to devise a set of ordering rules which might absorb these exaggerated fluctuations, instead of amplifying them. To do this, to see how the whole system reacted to fluctuations, a computer simulation was programmed to include typical fluctuations of demand and all the delay times in shipment for some 20 typical kinds of items—both slow-moving and fast-moving—at all levels of the net. Once the demand statistics and the ordering rules were specified, the computer would crank out a typical year's history of operation, with its occasional outages and its fluctuations in inventory at each level. When the rules then used and the supply sergeant's idiosyncrasies were put into the computer, a record very similar to actual records was obtained, with average inventories and percentages of outages corresponding well with actual experience.

Thus the model was checked out and could then be used to experiment with the results of changing rules. Since the computer turned out the equivalent of a year's experience in 15 minutes, a great deal of time (and money) was saved, over experimenting with the system itself. As a result some rules were suggested which the simulation indicated would reduce average inventory somewhat as well as the percentage of outages. The Army bought the idea. The first year of operation did bring a reduction of outages and a small reduction in average inventory. However, even a 1 percent reduction in a $5 billion inventory is quite a few millions. The system, with a few additional improvements that their analysts added, has now been in use for about 10 years.

Whole Blood Inventory

Another, more recent study may be of greater interest—that of the inventory of whole blood in a hospital.[2,3] Whole blood cannot be kept longer than 28 days; after that time it must be processed into its var-

[2] J. B. Jennings, *Inventory Control in Regional Blood Banking Systems,* Technical Report 53, M.I.T. Operations Research Center, (July 1970).

[3] J. B. Jennings, "Blood Bank Inventory Control," *Analysis of Public Systems,* A. W. Drake, R. L. Keeney, and P. M. Morse, (Eds.), M.I.T. Press, 1972.

ious components: plasma, platelets, and so on. These are valuable, but not so valuable as the whole blood. Here the balance must be made between the chance of not having blood of a given type on hand (shortage probability) and the chance of having the blood become useless because it is too old (outdating probability) while some other hospital is short of the same blood.

Most of the major hospitals in the Boston area collected about one-half of the blood supply they used, and stored it themselves. Their tendency (naturally) was to hoard their supply and their only gesture toward interhospital cooperation was to release possible oversupply when it was 20 to 25 days old. Since the smaller the system the larger the relative fluctuations, this arrangement produced frequent glut and famine. Often one hospital would be out of blood of one type while others had to outdate some of their supply. Clearly a more unified organization could reduce outdating as well as shortages. The crucial questions were: what would it cost, what sort of operating rules were required, and how much would be the expected reduction in outdating and shortage?

The Massachusetts Red Cross Blood Center was the organization with potential for setting up the cooperative system. It was already supplying nearly one-half of the whole blood used by the hospitals and, with its bloodmobiles, it could handle the necessary transport as well as the central storage required by a unified system. They asked the M.I.T. Operations Research Center to study the problem. John Jennings, a graduate student, decided to take it on as his Ph.D. thesis topic.

The first task, as usual, was to gather data on average use, statistics of fluctuations, present rules of operation, and opinions of the various operating personnel. On the basis of these data and knowledge of queuing theory, he began programming a simulation of a system of a central bank plus cooperating hospitals, each with its own inventory. His program had to include the fluctuations above and below each hospital's estimate of the next week's demand, delays and costs of transporting blood between hospitals or from the Center, as well as the flexibility to try out various rules of operation. Several different sets of rules were tried out: daily or weekly readjustment of all stocks or a system whereby stocks of different ages would be kept at predetermined levels, with replenishment shipments, above some minimum quantity, provided on request from the Center or from other hospitals with oversupplies of older blood.

The simulation soon demonstrated that daily readjustments involved excessive transport costs and weekly readjustment did not sufficiently reduce outdating and shortages. The more flexible system of replenishment-as-needed (according to a set of stock rules) did not cost too much for transport and gave promise of reduction of loss. The individual hospital stock rules determined the balance between the probabilities of shortage and of outdating, and this balance had to be decided by a consensus of opinion of the hospital staffs. The decision was to make the chance of outdating equal the chance of shortage for each type of blood. With these rules the simulation predicted that, with a system of five hospitals and the Center, both outdating and shortages could be reduced to about one-half the values experienced with the system (or lack of system) then in practice. The communication and clerical costs for keeping the requisite records were not excessive.

This system was installed and has been running now for several years. The savings predicted by the simulation have been achieved and all parties are pleased.

As with other cases, the details of all parts of the operation had to be known—what actually happened, not what one hoped would happen–before the simulation could be trusted to give correct predictions. In the case of the blood banks, stock rules were tried out at individual hospitals to see whether those elements went as predicted, before the system as a whole was adopted. In simulations involving proposed new equipment, such as new weapons, there is great danger that optimistic estimates will result in completely unrealistic predictions. Where possible, operational tests of the elements are the only safeguard.

Automobile Traffic

Computer simulation has also been used to analyze automobile traffic. One of the first such was carried out by Walter Helly for his Ph.D. thesis at M.I.T.[4] In single-lane, dense traffic, such as occurs in tunnels, the flow behaves like a peculiar sort of fluid. That is, any fluctuation of speed by one car generates a shock wave, traveling back from that car (often, traffic twenty cars back will have to come to a screeching stop). These shock waves will seriously diminish the rate of flow below that possible with a smooth motion. The Port of New York Authority had been experiencing this kind of trouble in its several traffic tunnels during weekend peaks of travel. They asked the M.I.T. Operations Research Center to study the problem and Helly took on the job.

[4] W. A. Helly, *Dynamics of Single-Lane Vehicular Traffic Flow,* Research Report No. 2, M.I.T. Operations Research Center, (October 1959).

The simulation had to include the interaction between successive cars in the stream. Experiments at the General Motors Research Laboratories showed that the usual driver preferred to keep a certain distance between himself and the car ahead—the greater the speed, the greater the distance. But in case of a change of speed of the head car, fluctuations would occur in this distance because of the limits on a car's acceleration or deceleration and also because of the delay in reaction time of the driver of the following car. Helly programmed an elaborate simulation of a single-lane flow, using all these data, with possible variabilities in reaction times, and other factors. After some readjustment of parameters and simplification of the model, it began to correspond fairly well with the rather sparse traffic-tunnel data. It indicated that the deceleration of heavy trucks, as they started on the upgrade, was often enough to start a shock wave.

A series of simulated experiments were started to see what could be done to improve matters. Some of the trials resulted in rear-end collisions—simulated on the computer, of course. Thus experiments could be carried out on the computer which would be too dangerous to carry out in "real life." Finally, it was determined that dividing the flow into "platoons" of about 30 cars each, with a space between platoons large enough to break most shock waves, would actually increase throughput by about 6 percent. This method was tried; and the reported increase was about 5 percent, which made it worth instituting during traffic peaks.

Application of Operations Research

All of these examples point the way to a number of generalizations about the application of the methods of physical science to operations. The first is that in any complex operating system, particularly those in the public sector, there are important aspects that are non-quantifiable (at least at present). Certain policy decisions must be made by the administrator on the basis of his experience. Therefore the operations analyst must work closely with this administrator, showing him what are some of the likely outcomes of his decisions and incorporating into the study the administrator's opinions regarding the choice of goals and the effects of politics and morale. Sometimes the boss must decide on a policy resulting in reduced efficiency, because of these other aspects. But at least he will know what is being sacrificed. This interplay between the analyst and the boss is one of the more difficult—but also more rewarding, if successful—parts of the task of the applier of operations research.

The question as to the goal of the operation is one of the most difficult aspects, particularly in the public sector, where profits are not the overriding criterion. In a traffic network of stoplights, parking rules, buses, and superhighways is the measure of success the greatest flow of cars, the quickest transport of people, or the least number of accidents? It may very well be that if you optimize one you may deoptimize the others. Should the health program reduce the number of working days lost from sickness, or should it concentrate on cases needing hospital care?

Often the first reaction of the administrator is in absolutes. We want the air over our cities to be as pure as it was in 1890. But, are we willing to agree to all the consequences of this program? The whole question of quantifying (as much as possible) the opinions of an experienced administrator is a part of the new field of decision analysis. Can one translate into mathematical terms his intuitive feelings about the proper balance between conflicting aims? Can the resulting combined objective function then be used to come up with an optimal solution?

In the public sector the final arbiter is the general public. A mayor may put into effect a policy he is convinced will improve things, but if the initial effects are not pleasant he may get voted out of office. So it is important to know what the public wants—not what they think they want, but what they will be satisfied with when it arrives. Public opinion surveys, of the sort we get in the papers, are not of much help. They usually ask questions which are too-general, or too-loaded. In general, they ask for an off-hand opinion, and do not delve deeper to find basic attitudes. This may be useful in regard to an opinion about a person, but they can be quite misleading when dealing with complex balances between opposing goals. Surveys too often ask for absolutes, when the proper balance between opposing action should be sought. Too many people think they want something, only to change their mind when reality is upon them.

This is a wide-open field at present. It could use trained psychologists and sociologists as well as experts in decision and marketing theories. One must watch what people do, in addition to asking them what they think they would do. Any improvement in estimating long-range public reaction would be of immense importance, for the determination of the proper measures of effectiveness is basic to all work in operations research in the public sector. A new project at M.I.T., called

Project Feedback, is an attempt in this direction. Its precursor was a study carried out by Stevens and Little for the Governor of Puerto Rico to find out how the very poor people of that island were feeling about the various attempts to improve their lot.[5]

Although we should recognize that this way of utilizing the methods of physical science has its severe limitations, we must also see that it has great potentialities, when carefully used.

BIBLIOGRAPHY

Ackoff, Russell L., and Maurice W. Sasieni, *Fundamentals of Operations Research,* Wiley, 1968.

Bellman, R., *Dynamic Programming,* Princeton University Press, 1957.

Cohen, Jacob W., *The Single Server Queue,* North Holland, 1969.

Cox., D. R., and H. D. Miller, *The Theory of Stochastic Processes,* Wiley, 1965.

Dantzig, George B., *Linear Programming and Extensions,* Princeton University Press, 1963.

Drake, A. W., R. L. Keeney, and P. M. Morse, (Eds.), *Analysis of Public Systems,* M.I.T. Press, 1972.

Jaiswal, N. K., *Priority Queues,* Academic Press, 1968.

Larson, R. C., *Urban Police Patrol Analysis,* M.I.T. Press, 1972.

Morse, P. M., and G. E. Kimball, *Methods of Operations Research,* Wiley, 1951.

Morse, Philip M., *Library Effectiveness: A Systems Approach,* The M.I.T. Press, 1968.

Morse, Philip M., *Queues, Inventories and Maintenance,* Wiley, 1958.

Morse, Philip M., and Laura W. Bacon, (Ed.), *Operations Research for Public Systems,* The M.I.T. Press, 1967.

Prabhu, Narahari U., *Queues and Inventories: A Study of Their Basic Stochastic Processes,* Wiley, 1965.

Takacs, Lajos, *Combinatorial Methods in the Theory of Stochastic Processes,* Wiley, 1967.

[5] J. D. C. Little, C. H. Stevens, and P. F. Tropp, "Puerto Rico's Citizen Feedback System," *Analysis of Public Systems,* A. W. Drake, R. L. Keeney, and P. M. Morse, (Eds.), M.I.T. Press, 1972.

7

Systems Engineering at the Jet Propulsion Laboratory

WILLIAM H. PICKERING

Director
Jet Propulsion Laboratory
California Institute of Technology

William H. Pickering is the Director of the Jet Propulsion Laboratory of the California Institute of Technology. He is a member of the National Academy of Sciences and a Charter Member of the National Academy of Engineering. Pickering has received the NASA Distinguished Service Medal, the Army Distinguished Civilian Service Medal, the Columbus Gold Medal, the Robert H. Goddard Memorial Trophy, the Crozier Gold Medal, the Spirit of St. Louis Medal, and the Italian Order of Merit.

The Evolution of JPL

I present here some thoughts based on practical experience with systems engineering at the Jet Propulsion Laboratory. Since 1940, JPL has developed from a graduate student thesis project in the aeronautics department of Caltech into the present organization with about 4000 members, 2000 of which are professional engineers and scientists working under contract to NASA to build and fly spacecraft to the planets. JPL has evolved from a purely research-oriented laboratory into one heavily engaged in the practical application of systems engineering of large and complex projects. In the process, we have developed for ourselves many of the principles leading to the successful application of systems engineering, and we have discovered how to organize an engineering team to accomplish a difficult project.

At the end of World War II, the Laboratory was working for the U.S. Army. It had been supported in the first years of the war by the Army Air Corps and given the task of understanding the principles of rocket motor design for application to aircraft problems. The Laboratory was successful in developing the jet-assisted takeoff (JATO) principle and was then asked to transfer its know-how to a commercial organization which would build JATO units in quantity (see Fig. 1). This phase of JPL's activity was essentially representative of engineering research. We were concerned with the engineering principles of the design of successful rocket motors, but very little with the problem of engineering these motors into an airplane, or of analyzing the application of these motors to an airborne mission.

Toward the end of the war, Army Ordnance asked the Laboratory to explore the application of rockets to long-range artillery applications. (In those days, "long range" meant 100 miles.) It was now necessary to understand a rocket system consisting of a rocket motor, fuel tanks, guidance, and payload. Accordingly, new types of engineers, such as aerodynamicists and electronic engineers, became part of the organization. The group began to function as a systems engineering team. Nevertheless, the emphasis was still on research and, because it was not yet necessary, very little systems engineering discipline emerged. While various engineering groups worked together on mutual problems and were responsible for many successful projects, by today's standards, they would be described as disorganized. The objective was not to optimize a design, but to build something that worked and to understand the engineering principles of the design. This, of course, is necessary before parametric studies leading to optimum solutions can be undertaken.

The next step in the evolution of JPL to a systems engineering organization was brought about by a request from Army Ordnance to develop an operational missile system. As there was a stated need to arrive at a production stage as soon as possible, it was agreed that, using the research rocket which had flown successfully, JPL would develop an operational missile system. This really meant that as many off-the-shelf items as possible would be put together to produce a workable device which, however, would be far from a well designed and engineered system. Indeed, this proved to be the case for this missile system, which was called the Corporal.

The system did work, and the military made it work even better, but it was expensive, inefficient, and required large amounts of support equipment. It pointed out the consequences of putting a system together rather than engineering the system.

From the Laboratory's point of view there were some valuable lessons to be learned:

- If a complex physical system is to be operated by relatively un-

Figure 1 The first jet-assisted takeoff (JATO) in the United States, August 6, 1941.

skilled personnel, the total system should be designed with a clear understanding of its end use and of the man-machine interface.

- The compromises inherent in bringing together devices designed for other purposes can only result in producing an inefficient system, difficult to operate and costly to buy and maintain.
- In order to assure an integrated and optimized system, design responsibility and authority for the complete system must be given to the implementing agency. There must be short communication channels between the various technical and engineering groups responsible for the development of the physical hardware and the project management which has the responsibility for attaining stated objectives. Then there can be reasonable assurance that, in optimizing the hardware to solve a technical problem, the overall

project objectives and constraints will be recognized and incorporated into the design.

- The transfer of knowledge from a developing agency to a producing industrial company is not simple. There are three major problems. First, it is practically impossible to document all of the important know-how involved in implementing complex hardware. It is too easy to fail to mention some procedure which is so much a part of your folklore that you just take it for granted. The second problem in transferring knowledge is that different organizations not only have different folklore but different methods of solving problems, of establishing acceptance criteria, of assuring quality. Third, the engineer who has developed the system finds that he must interpret his decisions to a group of people who have not been exposed to the effect that the system constraints have had on those decisions. Consequently the system documentation must be educational, as well as definitive, and the engineer will find himself involved in the educational process.

The Laboratory had an opportunity to show Army Ordnance that we could develop a better missile system than the Corporal, when we were asked to do the second generation Sergeant missile (see Fig. 2).

Figure 2 The Sergeant, solid propellant, tactical guided missile.

This assignment included the total system responsibility, with realistic user requirements and constraints. We developed the system, transferred it into industrial production, and assisted the Army in its initial field operations. Thus we were assigned a classical systems engineering task for a large and complex system. The project was successfully completed at just the time when we transferred to NASA in 1958.

Since joining NASA, the Laboratory has expanded its systems engineering capabilies. Our assignment for NASA is to conduct unmanned spacecraft missions to the moon and the planets. Hence we have developed and carried into practice a number of spacecraft systems which required advanced concepts and the welding together of many scientific and technical disciplines (see Fig. 3). Some of the systems considerations involved in these missions are discussed below.

Systems Concepts in Lunar and Planetary Projects

The capability for successfully accomplishing lunar and planetary missions was realized when five unique and relatively new space technologies were developed:

1. A launch vehicle for injecting a spacecraft into a trajectory designed to intercept the moon or a planet.
2. A spacecraft capable of operating unattended in space with a high order of reliability for periods of days or months.
3. An in-flight propulsion maneuver capability to provide the necessary target accuracy.
4. A system for determining the cislunar or interplanetary orbit of the spacecraft.
5. A system for communicating with the spacecraft: an uplink for commands and a downlink for data transmission and tracking.

These missions require systems consisting of perhaps a hundred thousand parts, several major systems contractors, many subcontractors, and projects requiring tens of thousands of man-years of effort. These projects often involve one- or two-of-a-kind designs, with cost and schedule constraints rarely permitting test flights. Thus the projects must be carried out in the presence of large unknowns, which include technical feasibility (always encountered in advanced designs) and the environments in which the systems must operate.

Systems Engineering

To carry out these projects with reasonable expectation of optimizing performance or of attaining project objectives within cost and

(a)

(b)

Figure 3 Four of the Spacecraft developed by JPL: *(a)* Lunar Ranger VII, *(b)* Mariner Venus, 1967, *(c)* Mariner Mars, 1969, and *(d)* Mariner Mars Orbiter, 1971.

schedule, a systems engineering approach is obviously necessary. Basically, the systems approach involves the optimization of the overall system as opposed to the piecemeal suboptimization of the elements of the system. This overall optimization is achieved in a number of steps:

1. Goal definition or problem statement.
2. Objective and criteria development.
3. Systems synthesis.
4. Systems analysis.
5. Systems selection.
6. Systems implementation.

The systems engineering of lunar and planetary missions—conceptualization, design, fabrication, operation—is carried out according to this scheme.

The goals of lunar and planetary programs are formulated in terms of a sequence of lunar and planetary missions, with objectives for each mission based on scientific and engineering feasibility studies. Design criteria provide a means of ranking alternative systems, establish the importance of various phases and objectives of the mission, and furnish a basis for making trade-off studies.

The systems synthesis and analysis is initially carried out in terms of functions required for the fulfillment of the mission. The design is then successively iterated for the effects of various hardware implementations. The selected design represents an optimum choice with respect to the mission objectives and the design criteria.

To focus and integrate the management and engineering efforts for a specific mission, a project organization is needed. Support from the total technical resources of the organization is provided by a "matrix" organizational structure. The classical organization chart with vertically aligned functional divisions is overlaid with horizontally aligned projects which intersect the division structure. An engineer from a functional division, working on a project, administratively reports to his functional division, but receives his work assignments from the project management. His performance is jointly reviewed by both the division and the project.

These projects must be broken down into comprehensible parts in order to be accomplished. The first major division of a space project is into systems. Lunar and planetary flight projects typically are composed of four systems (see Fig 4):

1. The Spacecraft System, consisting of the spacecraft and its support equipment.

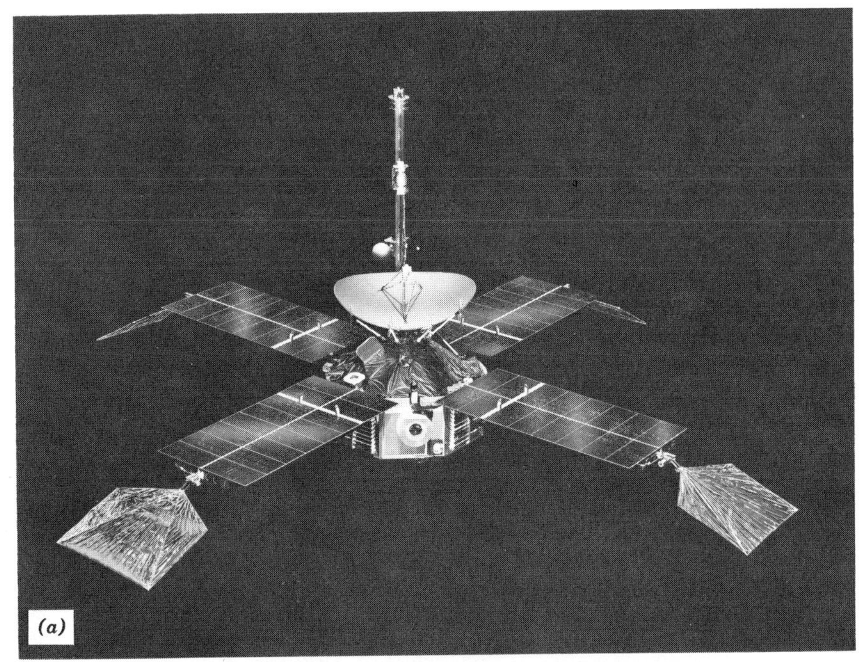

2. The Launch Vehicle System, consisting of the launch vehicle and its support equipment.
3. The Tracking and Data System, which is responsible for the provision and maintenance of the earth-based tracking, telemetry, and command stations; the ground communications; and the operational facilities for the mission.
4. The Mission Operations System, comprising the management organization responsible for the design and execution of the mission operations.

Further breakdown of the systems into subsystems and components is made to reach a level of complexity that can be treated as a single element. Concurrent with the system breakdown, interfaces between the elements are established which define the functional boundaries of the elements.

This breakdown of a system into functional elements cannot be made arbitrarily. A great amount of managerial and engineering skill is required to select the interface topology, which affects the management control of a project, both administratively and contractually, and the engineering integration and operation of the system. From a management standpoint, interfaces are defined so as to optimize

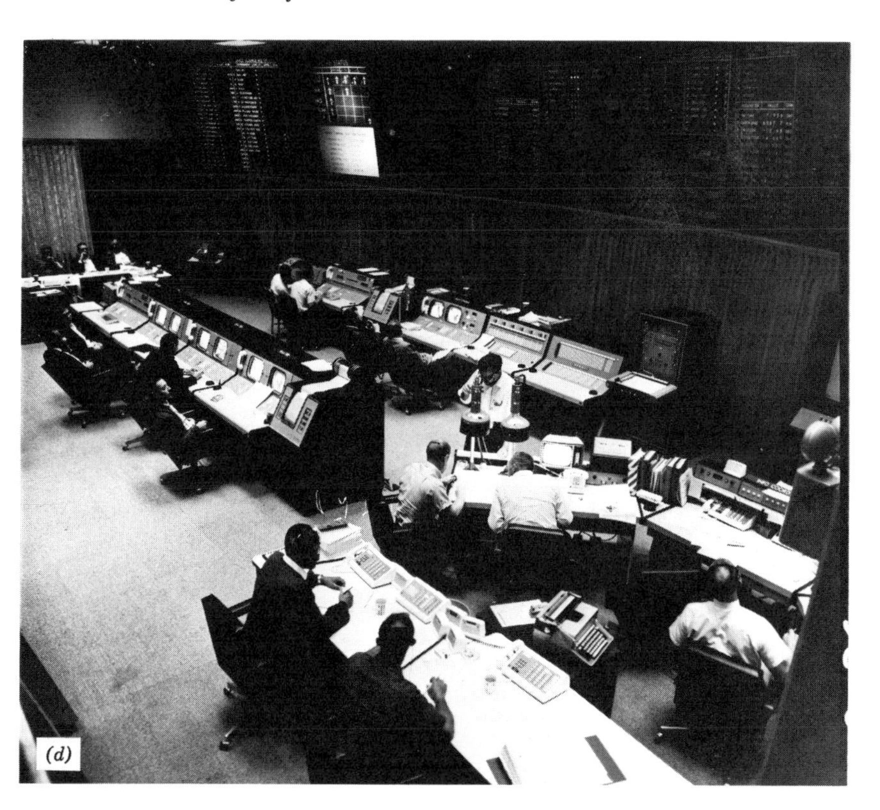

Figure 4 The four systems of a lunar or planetary flight project: *(a)* spacecraft system, *(b)* launch vehicle system, *(c)* tracking and data system, and *(d)* mission operations system.

visibility and control, to isolate independently subcontracted elements, and to delegate authority and responsibility. From an engineering standpoint, interfaces are located so as to separate independent functions and to facilitate the integration, testing, and operation of the overall system.

In the early design stages, the exact mode of implementation for a system may not be known. For example, should the spacecraft be spin stabilized or inertially stabilized? Should power be obtained from solar cells or a nuclear device? Thus a system is initially defined in terms of performance specifications and constraints for each function, rather than in terms of the implementation schemes.

At this point, the design is defined by a set of subsystems, functions, and constraints for each subsystem, and an outline of the interface topology. Then the subsystem designs are projected to the point that the alternative subsystem implementations are understood. The subsystem performance characteristics and constraints are translated back through the system, and the process is iterated to produce an optimized, self-consistent preliminary design.

The preliminary design is specified as a set of functional requirements and interface control documents. These documents define the overall requirements and constraints levied on the spacecraft design by the mission, the major system interfaces, the subsystem interface topology, and the functions and constraints imposed by the system on the subsystem designs. They describe the design in sufficient depth to allow the detailed subsystem definition to proceed independently.

Design for Success

Reliability. Perhaps the most difficult requirement to satisfy in space missions is reliability. It presents the most challenging criteria for the management and engineering of deep-space systems.

There are typically tens of thousands of electronic parts in a planetary spacecraft design, many in themselves miniature assemblies (see Fig. 5). In the early 1960s, there was a serious concern that the sheer number of parts, each with its small but finite probability of failure, would result in an intrinsic unreliability at the system level that would render planetary flights unfeasible. Failures on early space flights tended to add credence to this concern. It took a fundamental change in the implementation of space projects before mission success occurred with regularity. It is now clear that the process of achieving reliability must start with the first concept of how a mission will be implemented and does not end until the last mission event has been successfully completed.

A Conservative Approach. It is necessary to take a very conservative approach to the design of lunar and planetary spacecraft in order to obtain the levels of ultrareliability required to achieve a high probability of mission success. The most conservative designs capable of fulfilling the mission requirements must be considered. This involves, wherever possible, the use of flight-proven hardware and, for new designs, the application of state-of-the-art technology, thereby minimizing the number of unknowns present in the design. New designs and new technologies are utilized, but only when already existing flight-proven designs cannot satisfy the mission requirements, and only when the new designs have been extensively tested on the ground.

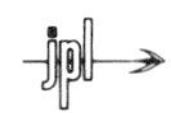

SPACECRAFT PART COUNT COMPARISON

PART CLASS	MARINER IV	MARINER V	MARINER 1969	MARINER 1971
CAPACITORS	5,570	4,594	3,222	3,957
RESISTORS	15,607	10,781	9,916	11,031
DIODES	9,922	5,047	4,418	4,748
TRANSISTORS	4,323	3,027	3,035	3,296
IC's ACTUAL	–	594	2,763	3,063
(IC EQUIVALENT)	–	(17,226)	(80,127)	(88,827)
MISC*	3,798	1,225	1,105	1,256
ACTUAL TOTAL	39,220	25,268	24,459	27,351
(EQUIVALENT TOTAL)	39,220	(42,494)	(104,586)	(116,178)

*INCLUDES SWITCHES, CRYSTALS, FUSES, INDUCTORS AND TRANSFORMERS, RELAYS, ETC.

Figure 5 Spacecraft electronic parts count comparison.

Simplicity. Simple designs tend to minimize the number of parts, functional modes and interfaces, and have the fewest unknowns in terms of interactions with the system and with system operations. The philosophy of a fail-safe design should also be used; the spacecraft should be capable of performing a major portion of the mission even in the presence of failures. The system should have the capacity for operating in spite of some degradation in major parameters.

Redundancy. An important method for obtaining reliability at the system level is with redundancy. All mission-oriented functions should be backed-up by redundancies or alternate modes, and the system should be protected against failures in noncritical elements. Communications and power subsystems, being mission-critical elements, typically contain some element redundancy. On Mariner 1969, most of the computer events were backed-up by ground commands. In addition, there were more than a dozen cases of element redundancy.

Interface design. This is an extremely important part of the overall mission design. Interfaces exist between elements of the system, between systems, between people, and between organizations. The tranfer of functions across interfaces tends to be particularly susceptible to design error. Hardware interfaces often also involve responsibility interfaces between different organizations. Thus it is essential that interface designs be simple in order that each side of the interface can be adequately designed and tested prior to mating, and in order that interface responsibilities can be clearly understood. On early Mariner spacecraft, the Spacecraft System/Launch Vehicle System interface occurred at the in-flight separation joint. On the Mariner 1969 spacecraft, the interface was moved down to the field joint. This eliminated the in-flight separation mechanism from the interface definition, thus simplifying the management interface as well as the testing and operations.

The design must also consider many other factors. To achieve the highest levels of confidence, the design must be capable of being tested and analyzed. Design confidence is obtained when test results correspond to analysis predictions.

Testing

The test program must be thoroughly integrated into the fabrication and assembly operations. The test results must be factored into the system design, and changes must be made where the test results indicate that mission objectives will be compromised.

The test program accomplishes many objectives. It provides data where additional information is needed to complete the design (development testing); it qualifies the design (type-approval testing); it validates the flight hardware for the mission (flight-acceptance testing); it produces calibration and signature data; and it provides training for the mission personnel (mission simulation). Testing, starting with the parts and materials, involves every level of assembly: components, subsystems, and systems.

The finite time available for testing restricts the range of operating states that can be investigated. The Mariner 1969 spacecraft with more than 10^{21} distinguishable states at the system level, even if sequenced through system states at the rate of one thousand states per second, could not be completely tested within the observed lifetime of the universe!

Mission Plan

Finally, the preparation for the mission is not complete until a detailed mission plan has been developed and tested, and the operations personnel have been trained in their duties. The mission plan must include an integrated, step-by-step account of the functions of all systems, through all phases of the mission. The mission plan should maximize the mission return in a reliable manner, consistent with the mission objectives and the constraints of the project systems.

Project Management

The transient nature of projects, the fact that no two projects employ the same resources, and the complex nature of the missions require that initially the mission objectives and the resources allocated to a project be specifically identified. Responsibility and authority must also be clearly delineated.

Lunar and planetary projects at JPL are managed in accordance with a management procedure specified in a Project Development Plan (PDP), which is prepared under guidelines provided in NASA management instructions. When management responsibility is assigned to JPL, the project management reports administratively to the Flight Projects Office at JPL, and technically both to it and to a counterpart program office within the NASA Office of Space Science.

Basically, the organization of JPL is focused on technical or professional disciplines. The majority of personnel and groups supporting project efforts within the Laboratory are members of various technical divisions. Each of these divisions assigns a full-time division representative to the project to assume responsibility for the efforts of the division for the project. Virtually all divisions, including service and support elements, participate to some degree in the activities of the project.

Project management must continually ask four questions: (1) will it work? (reliability), (2) will it operate as specified? (performance), (3) will it be ready? (schedule), and (4) for what cost? (resources). The responses to these questions concerning reliability, performance, schedule, and resources form the information base for a continual assessment of the status and progress of the project.

Project management interacts with the project elements through scheduled working meetings, which are conducted from the project initiation through to the conclusion of the mission. The project management interacts with NASA Headquarters, JPL management, and the project elements through a series of design reviews. Preliminary, design, and hardware acceptance reviews are conducted at system and subsystem levels. These reviews consist of presentations to a board, usually supported by back-up documentation in depth, followed by recommendations forwarded by the board to the project or concerned system manager. The general purpose of the reviews is threefold: (1) to bring independent and senior judgment, in the form of the review board, to bear on all aspects of the system or subsystem; (2) to assure consideration of the internal and interface characteristics by the appropriate managers and engineers; (3) to uncover and respond to residual or new problems.

The project attains its performance and reliability goals through the integrated stages of assembly, test, and assessment—starting with the procurement of parts and materials and culminating in the launch and encounter readiness reviews. The system performance is assessed by comparing test and analysis results against performance specifications. The reliability goals are achieved at the project level through an extensive reliability program involving many aspects of the project efforts.

To attain the reliability goals, there must be a parts qualification and control program, and the hardware fabrication process must be controlled. A quality assurance program must control hardware workmanship and assure that all test objectives have been met. There must be a configuration management program capable of identifying and verifying the specific components assigned to each spacecraft. An integrated test program is necessary, starting with the parts and materiel and providing functional and environmental tests of system elements at all levels of system assembly.

In addition, there must be a controlled problem/failure identification and resolution system capable of answering the following questions: (1) What failed? (2) How did it fail? (3) Why did it fail? (4) How was it fixed? (5) Why won't it happen again?

Finally, there must be a detailed mission plan, to be executed by qualified and trained personnel, with systems which are to be operated within tested and analyzed envelopes.

The timely management of project costs and schedule is extremely important, because unexpected problems invariably result in increased costs and schedule slippages in the absence of corrective action by project management. Project status must be continually reassessed: actual costs must be compared against planned costs, and progress against schedule.

Schedule control is absolutely essential to the successful completion of planetary missions. While it is possible to slip lunar flights on a month-by-month basis, opportunities for planetary launches occur at widely separated intervals—19 months between launch periods for Venus flights and 25 months for Mars flights. Thus, project problems must be identified and resolved without schedule slippage.

A control similar to the Mariner 1969 schedule milestone chart shown in Fig. 6 is used on most projects for summary reporting. It is not intended to be used for detailed planning because it may not

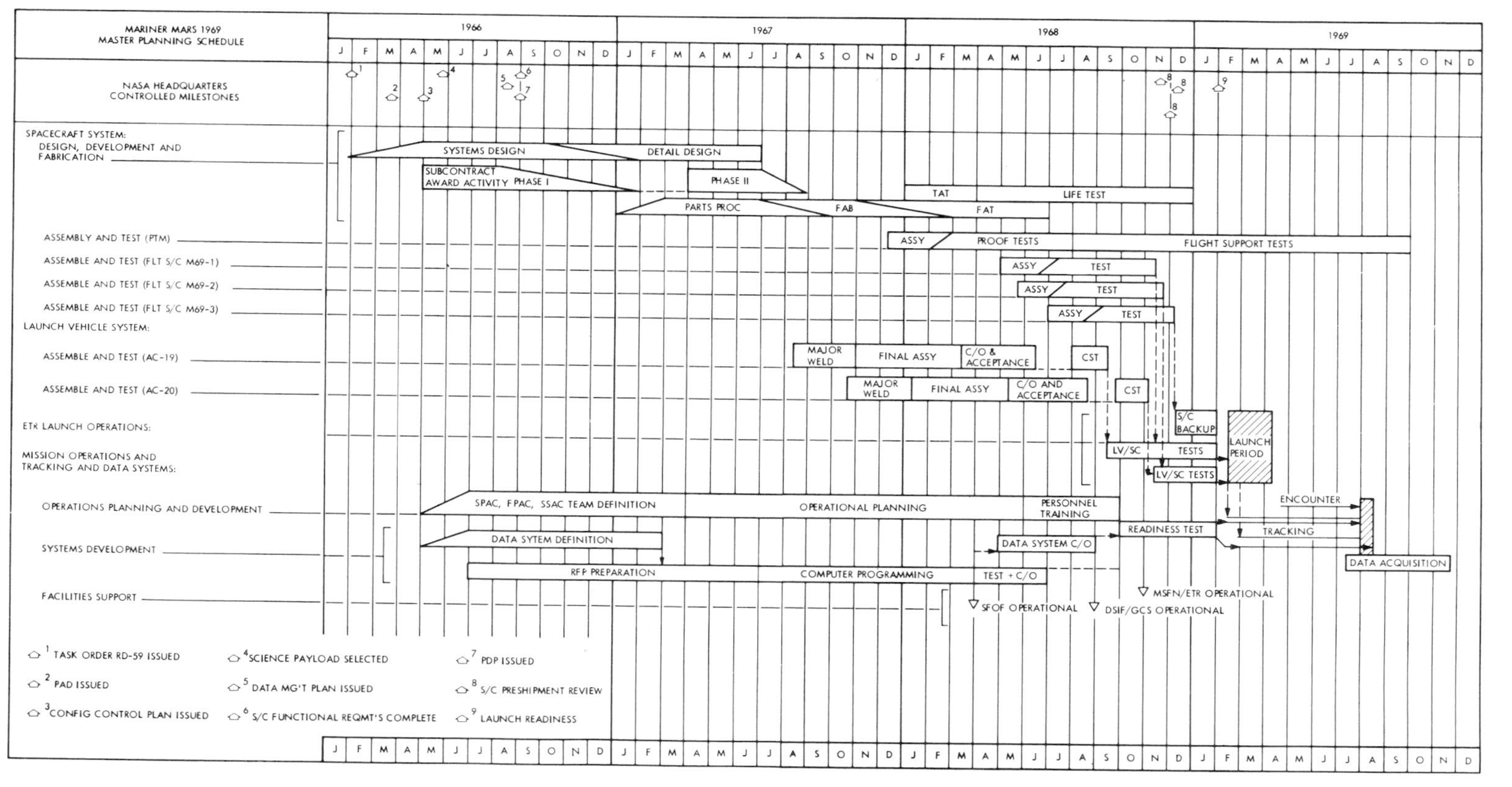

Figure 6 Mariner 1969 project master schedule.

permit early detection of slippages, and it does not show the interrelation between the scheduled activities.

For this purpose, the Navy, in the late 1950s, developed PERT: Program Evaluation and Review Technique. Through the display of events, activities, and their interrelationships, PERT is a tool by which project management can estimate their effects, and select the "critical path" of those activities which cannot be delayed without affecting project schedules.

Cost is always a hard constraint. It is necessary to size a project to the available resources, and to redirect the project's efforts when the cost constraints are approached. In addition to keeping an accurate up-to-date account of present and past project expenditures, it is necessary to continually extrapolate the integrated project costs out to the completion of the mission.

Figure 7 shows the history of the estimated cost-to-completion for the Mariner Mars 1969 Project. The early increases were associated with the selection of the scientific payload and the decision by NASA to proceed with that payload, and the later increases were associated with the progressive realization of subsystem contractor costs and spacecraft change costs.

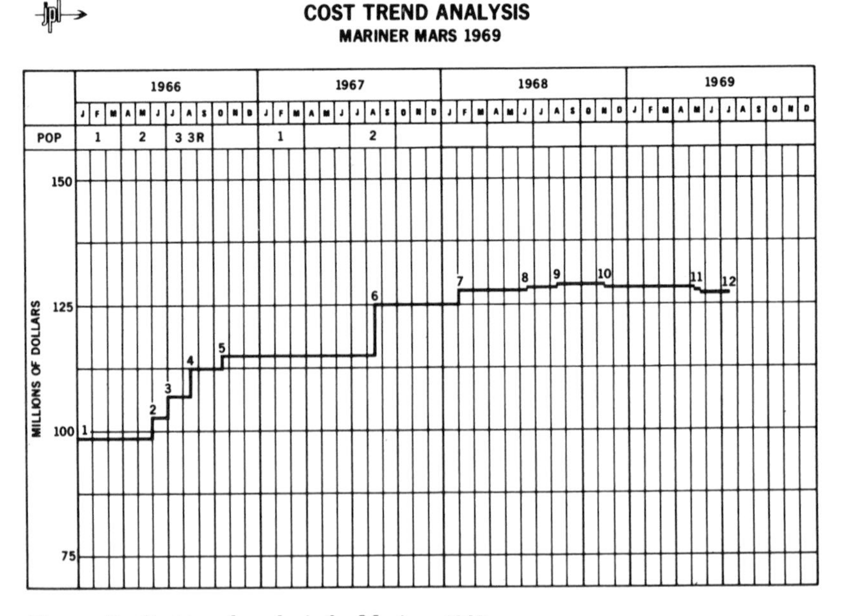

Figure 7 Cost trend analysis for Mariner 1969.

Civil Systems Projects

By 1965, it had become apparent that space-associated technologies and systems engineering techniques existing at JPL could be productively applied to other activities in the national interest.

Morgantown Project

Recently, JPL completed the initial design phase of a Department of Transportation project to demonstrate the feasibility of a new systems concept of public transportation responsive to the needs of a large university in an urban community (see Fig. 8). Involving West Virginia University, the city of Morgantown, and the Urban Mass Transit Administration (UMTA) of the Department of Transportation, this project is intended to demonstrate the technological, operational, and economic feasibility of a computer-controlled, fully automated transportation system in demand-activated and schedule-service modes, and to demonstrate its compatibility with other forms of transportation. The system will use a fleet of computer-controlled vehicles operating on a dedicated guideway between six off-line stations along a 3.5 mile route.

Under the overall direction of UMTA, JPL had the systems management responsibility for the initial design phase of the Morgantown Project. The systems management responsibility has now reverted to UMTA, which is presently implementing the system with contractor support. The West Virginia University has assumed the system-user management responsibility, including the coordination of the University and the community requirements, and is preparing for the system operations and maintenance.

During the initial design phase, the Morgantown Project at JPL was directed by a JPL project manager who reported to the Laboratory Director's Office. The JPL project line and technical organization was divided into four major areas: Vehicle System, Control and Communications System, Structures and Power Distribution System, and Project Integration. The system manager of each of the four areas was assisted by a team appointed from the JPL technical divisions. JPL management continually evaluated project progress through committees and review boards.

On-site guideway construction has begun. Subsystem and system testing began in 1972, with a prototype demonstration scheduled to continue for one year. The prototype section of the system consists of three operational stations with connecting guideway, 5 prototype vehicles, and a control center. A formal Joint Acceptance Test in

Figure 8 Artist's concept of Morgantown people-mover project.

October 1973 will verify that the primary project requirements have been met and that the system is ready for public use.

The final phase of project testing will demonstrate that the system is operationally satisfactory and can meet actual traffic demands. The first evaluation will be complete by February 1974, the second in October of that year. During this period, the transition to the University will begin and should be complete by November 1974.

System description. Vehicles will be fully automatic, with no on-board operators. Automated functions include control of vehicle speeds for minimum safe headways, position verification, and performance monitoring to determine degraded conditions. The vehicles will be heated, air conditioned, and powered electrically from a third-rail pickup. Each car receives its operating instructions through an inductance loop and cable. The cars will run on four single, rubber-tired wheels. An on-board device will be used to steer the vehicle, contacting a sensor to follow the guideway geometry.

Nominal demand-mode platform waiting times will not exceed 2 minutes; in the schedule mode, not over 1/2 minute. Upon leaving the platform, the vehicle will move slowly to a dispatch point and await a computer command to accelerate before merging into the main guideway traffic. On the guideway, vehicles will be maintained at fixed distances and speeds of 30 mph. As the vehicle nears its destination, it will be commanded to exit at the next ramp and decelerate as it approaches the platform track merge point.

The guideway will consist of fixed running surfaces permitting one-way traffic flow in each direction. Access trackage will connect the main guideway with the stations and the maintenance facility. The guideways will be capable of essentially all-weather operations, with heaters below the running surfaces to prevent accumulation of snow or ice.

A central computer will control all operations, including vehicle start, stop, speed, loading, unloading, switching, and all monitoring functions. In the normal mode of operations, all of the control signals are generated automatically at the control center computer as functions of stored programs responsive to either preprogrammed (schedule) or real-time (demand) data inputs.

The Morgantown project is truly an interesting and significant one, pioneering in the innovative application of systems engineering and management techniques to civil-sector programs.

Urban Health Systems Task

Another JPL civil systems project, the Urban Health Systems Task, involves identification and implementation of methods for the improvement of the quality of health services in the urban community of South-Central Los Angeles, including Watts.

Despite years of remarkable medical achievement, the health delivery system in this country has remained relatively unchanged, although there is great need for innovative forms of organization equal to the task of applying new techniques and knowledge. In 1970 the Los Angeles County Health Service Planning Committee proposed that the County should adopt a health care delivery system integrating hospitals, diagnostic clinics, ambulatory care centers, and home-care programs into a community based network. South-Central Los Angeles was selected as the first area of the County in which such a network would be established. The newly constructed Martin Luther King, Jr. General Hospital was the first County hospital to be integrated into the network.

The Urban Health Systems Task is a cooperative effort between JPL and the Martin Luther King, Jr. General Hospital. Recently completed was a systems analysis of an integrated health care delivery system for the area, similar to the mission analysis and engineering studies undertaken by JPL in early space project planning.

Specific objectives of this preliminary phase of the task were:

1. To determine the community health care needs.
2. To formulate criteria for evaluating the health care delivery system.
3. To develop a complete system description.
4. To demonstrate the feasibility and value of the more extensive efforts planned for succeeding phases by analyzing in detail three functions: emergency care, maternal and infant care, and a health-testing program.

The system description which was developed includes the elements of the system, the functions of the system and its elements, and the interactions that exist between the system elements. The elements of the system consist of the facilities, the organizations, the people, and the resources. The categories of people include the patients, doctors, nurses, technicians, and the administrators and students. The system resources are obtained from federal, state, and county funds, and from foundations and volunteer groups. A partial list of the major system-level functions includes preventative medicine, therapeutic medicine, education, administration, and facility maintenance.

A preliminary model was developed for the hospital admitting system, which controls patient flow. The required medical services for an arriving patient are typically grouped as obstetrics-gynecology, surgery, internal medicine, or pediatrics. This admitting system can be treated as a mathematical model, with probability distributions for the holding times at each station determined from available statistics of existing systems. It is possible to exercise this model, either by using a Monte Carlo computation process, or a General-Purpose Systems Simulator computer program. In this manner, a range of possible admitting system configurations can be investigated.

An analysis was made concerning specific recommendations on effective ways to improve the delivery of care in the three areas selected for detailed study. These studies included the following steps:

1. A description of existing and planned approaches.
2. An identification of the constraints on the existing care and a development of criteria for evaluating potential improvements.
3. An analysis of existing and potential resources.
4. An identification of specific quantitative performance requirements.
5. The development of recommended courses of action.

Conclusion

Thus we see that over the years JPL has worked on a variety of projects requiring the implementation of large hardware and software systems. The successful accomplishment of these projects needed an extensive application of the methods of systems engineering, and now we are hoping to apply these methods on the broader scale of civil systems.

Of course all engineering tasks have an element of systems engineering in them. If you propose to build a better mousetrap, you must consider many factors before you can pick an optimum and successful design. The term systems engineering, or systems approach, is, however, usually reserved for large engineering tasks which require the skills of a team of engineers encompassing many disciplines to accomplish the design.

The systems approach has been used successfully in many Department of Defense and NASA projects, and it is now being proposed as a means of solving many vexing problems in the civil sector. However, most of these cannot be solved by systems engineering unless many other factors are changed. To illustrate the point, let us consider the requirements on a project which is amenable to a systems approach:

1. The project must be defined. Before anything else can be done there must be a clear statement of the expected end result. For example, in 1961 the Apollo project objective was stated: send a man to land on the moon and return safely before the end of the decade.
2. The project constraints must be clearly understood. These will consist of costs and schedules and, equally important, of the numerous management, legal, political and other factors which limit the type of solution which is acceptable.
3. The project organization must be agreed upon. The hierarchy of responsibility and authority from the original source of project assignment and fiscal assignment down to the technician on the test bench must be clearly spelled out, and must be inviolate for the life of the project.
4. The project must be implemented by a team possessing the necessary skills and facilities. The necessary technology must be available.

Unless these four conditions can be satisfied, the systems approach cannot lead to a useful solution. In any real problem an additional criterion is needed. That is the values and risks associated with a given solution. If these can be quantified, then a mathematically optimum solution can be found. In practice it may be quite difficult to come up with any precise answers. Then the judgement of the project management must be counted upon.

In the case of civil systems it may be difficult to obtain definitive answers to any of the four criteria, and value judgments will certainly be difficult to obtain agreement upon. For example, suppose we wished to establish a project to eliminate or, at least, minimize smog in the Los Angeles basin. Even the project definition may be difficult. Research would need to be done to understand the factors affecting visibility, agriculture, and human comfort. Then agreement would need to be obtained as to which factors would be reduced by how much. As far as constraints are concerned, there are certainly many legal and political factors which would have to be considered, and there is also the question of acceptance of the solution by the citizens. To do the project, the organization would require clarification of authority. For instance, could the project management enforce a solution on the citizens or would it have to pass through state or federal legislatures? With regard to implementation, if the other factors could be settled, it would probably be possible to assemble the appropriate implementation organization. The final question of values could well be the most difficult to settle. Is it worth $10 or $100 per person to eliminate smog? This question could only be answered by going to the polls. Problems such as these only make clear that the systems approach, no matter what its theoretical merits, will be very difficult to apply to many civil systems.

Returning to JPL systems experience, note that most of the discussion has been concerned with systems management, problems of definitions of objectives and understanding project constraints, rather than on the application of mathematical techniques to solve the optimization problem. These techniques are used at JPL, but as stated earlier, the quantification of many of the key functions in the optimization equations is frequently very difficult and, therefore, judgment factors have to be employed. Parametric and optimization studies are carried out where possible, and indeed they are essential for some system decisions, as for example the choice of launch date for a planetary mission. However, many decisions must be made on little or no quantitative information. These are good decisions when the systems organization clearly defines the responsibility for the decision and when the systems team is functioning together as a unit and is made up of knowledgeable individuals.

The practice of systems engineering is, therefore, more concerned with management and with information than with mathematics. Subsystems on the other hand must be analyzed from the point of view of all performance parameters, and various mathematical and computational techniques are obviously needed.

The JPL venture into civil systems presents us with the opportunity to broaden our capabilities into these new areas and to understand the problems associated with the systems approach to such systems. The two systems, Morgantown and the hospital, present quite different problems. Morgantown is much closer to the spacecraft projects with which we are familiar, than is the hospital project. At Morgantown we have a project with definable goals and constraints. The end product is a hardware system which will be successful when it demonstrates its ability to transfer people at the rates and with the reliability specified. It must also be economical and it must be accepted by the customer. The project responsibilities are defined and the roles of the local city and university governments are understood. In sum, the Morgantown Project can be considered analogous to a spacecraft project, except that it needs some different technical disciplines and the man-machine interface is much more important.

The hospital project, on the other hand, is much less in the style of a conventional project. The objectives can only be defined in broad terms. The constraints are generated by interfaces with a multitude of agencies, the county government, the AMA, legal requirements, and so on. Responsibilities in the project are equally nebulous and are dependent upon the same interfaces. The project is primarily a software project and the results will be more difficult to evaluate.

BIBLIOGRAPHY

Corliss, W. R., *Space Probes and Planetary Exploration,* D. Van Nostrand, 1965.

James, Jack N., et al., *Mariner IV Mission to Mars,* Technical Report No. 32-782, Jet Propulsion Laboratory, California Institute of Technology, September 15, 1965.

Mariner Mission to Venus, Jet Propulsion Laboratory, McGraw-Hill, 1963.

Pickering, William H., "The Grand Tour," *American Scientist,* **58,** No. 2, (March-April 1970), 148-155.

Pickering, William H., et al., "TOPS—Outer Planet Spacecraft," *Astronautics and Aeronautics,* **8,** No. 9, (September 1970), Special Section.

Schurmeier, H. M., R. L. Heacock, and A. E. Wolfe, "The Ranger Missions to the Moon," *Scientific American,* **214,** No. 1, (January 1966), 52-67.

Systems Engineering in Space Exploration, Jet Propulsion Laboratory, 1965.

Urban Health System Report, JPL Document 650-148, Vol. I-VI, Jet Propulsion Laboratory, California Institute of Technology, to be published.

Wilson, James H., *Return to Venus (Mariner Venus 1967),* Technical Memorandum 33-393, Jet Propulsion Laboratory, California Institute of Technology, 1968.

Wilson, James H., *Two Over Mars: Mariner VI and Mariner VII,* NASA EP-90, U. S. Government Printing Office.

8

Apollo: Looking Back

GEORGE E. MUELLER

President
System Development Corporation

George E. Mueller is the President of System Development Corporation, a member of the National Academy of Engineering, and the Vice President of the International Astronautical Federation. From 1963 through 1969, he was the Associate Administrator for Manned Space Flight for NASA, and as such was responsible for all aspects of the manned space program. In 1970, he received the National Medal of Science for his contributions to the design of the Apollo System. Before assuming his present position, he was the Senior Vice President of General Dynamics Corporation. Mueller is a coauthor of *Communication Satellites*.

The engineer is a craftsman—in one of Webster's definitions, "a person who carries through an enterprise by skillful or artful contrivance." This definition would certainly apply to those who have managed and engineered the Apollo effort to explore the moon. In this chapter, rather than describe the formal procedures and the chronology of the Apollo program (all of which have been well documented), I try, through some recollections of my own experience in managing the program, to suggest the necessity for creativity and innovation in modern systems engineering. In spite of the massive scale of the Apollo program, in terms of both the resources applied and the technologies used, the major issues were ultimately resolved by a small number of people making difficult decisions in the presence of great uncertainties and unprecedented technical complexity (see Fig. 1).

The Apollo Budget

The Apollo program is the largest engineering enterprise ever undertaken. Before it is finished, it will involve the expenditure of $20 billion and the efforts of more than 200,000 people. I might point out here at the beginning that the $20-billion figure was arrived at in a very Spartan manner. Congress asked Hugh Dryden, NASA's Deputy

Figure 1 Apollo liftoff. (NASA.)

Administrator, how much it would cost to go to the moon and was told the cost would be between $20 billion and $40 billion. As one learns in working for the government, smaller figures are heard with far greater clarity than larger ones, and the Apollo budget was set at $20 billion. That amount was reviewed annually, and when I arrived in Washington to manage the program, it had been cut for the following year by $1 billion. My first experience with the program, therefore, was the sobering one of searching for things that were not absolutely necessary and cutting them out. This is a most valuable discipline in systems engineering.

The Radiation Hazard

Money, however, was only one problem; there were others, beginning with the fact that the scientific community was not of one mind either about the reasons for going to the moon or whether it was possible to get there. Both the President's Scientific Advisory Committee and Congress were raising serious questions about a variety of problems, one of which was the exposure of humans to radiation in space. The Van Allen belts, bands of rapidly moving charged particles circling the earth, had been discovered only a few years earlier, and no one was certain of their extent or the intensity of radiation associated with them. Also, solar storms had recently been discovered that inundated the space around the sun with avalanches of charged particles, producing radiation levels that would be lethal for a human in an unprotected state. Congress had heard about these things and wanted to know how we could propose to send men into space to sustain unknown and possibly fatal doses of radiation.

This was a difficult question. There was not much knowledge about either the frequency or the intensity of solar storms, and the Van Allen belts had not been explored thoroughly enough to indicate how far out they extended. We spent several months, therefore, intensively studying the levels of radiation between the earth and the moon. We were able to establish the upper limit (at least, the upper observed limit) of the intensity of solar storms, and we were able to show that, with some rearranging of equipment inside the cabin, the Apollo Command Module (see Fig. 2) provided shielding thick enough so that the men inside it could sustain the entire period of a solar storm without being exposed to excessive radiation.

The Lunar Module (LM) (see Fig. 3) presented a different problem because it had to be built of thin and light material that could not protect the astronauts from solar radiation on the lunar surface. (The space suits also offered practically no protection.) We determined how much lead shielding we would have to add to the LM to provide the necessary protection, and found that it would have made it too heavy for launching. What we finally did was to take advantage of the fact that solar storms build up gradually and that, although the radiation can become intense, it takes an exposure of several hours to do any significant damage. The problem then became one of detecting the onset of a solar storm, so we developed a network of sensors aboard the Command Module that would detect a radiation buildup; and we modified the mission plan in such a way that, if the radiation were to build up at a certain rate during the moon walk, the astronauts would enter the LM, take off, return to the Command Module, and abort the remainder of the mission. This was an indirect solution, but one which demonstrated that we understood the problem and had a plan to deal with it—the primary considerations.

The Meteoroid Hazard

Another group of questions had to do with meteoroids. Harvard-Smithsonian's Baker-Nunn cameras were recording thousands of what appeared to be meteoroid flashes in the earth's atmosphere every day. If we extrapolated from these recordings to estimate the density of meteoroids in space, it appeared that an ordinary space vehicle orbiting the earth for a week or two would be bombarded with mete-

Figure 2 The Apollo Command and Service Modules. (NASA.)

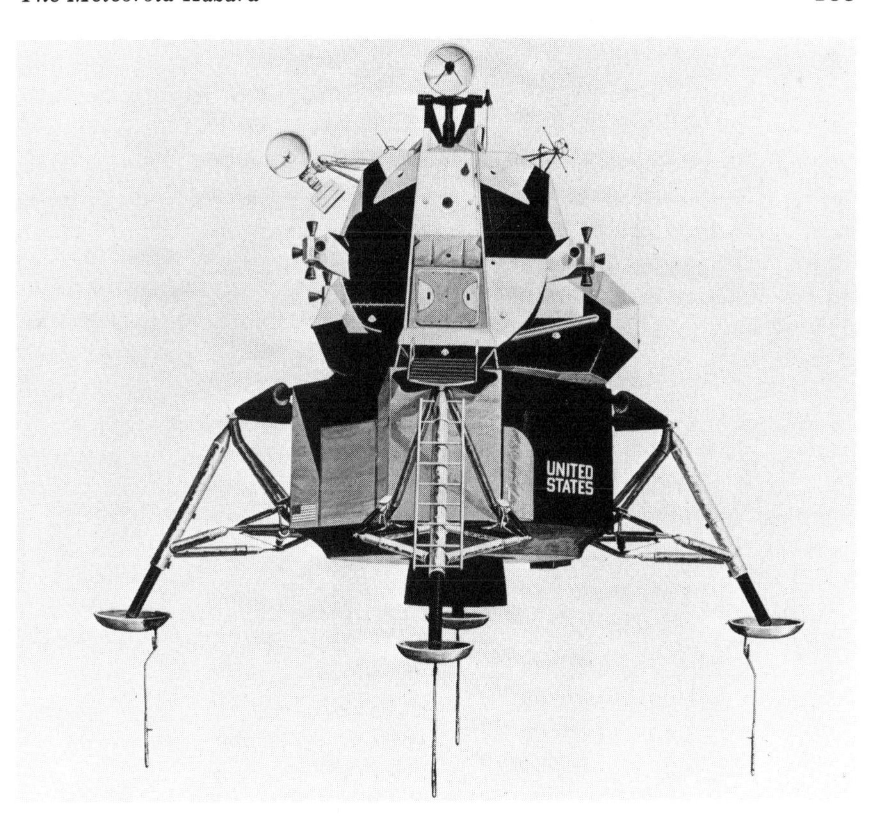

Figure 3 The Apollo Lunar Module. (Grumman Aerospace Corp.)

oroids and probably be damaged or destroyed. This data was discouraging, but then we got contradictory data from small satellites that were equipped with microphones; all they recorded was an occasional ping. At that point, no one knew what to believe. Finally, to try to resolve this dilemma, we developed the Pegasus satellite. Pegasus had large panels, about 10 ft. high and about 60 ft. long, that, when they unfolded in space, presented a large enough area to collect a reasonably good statistical sample of the number of particle impacts. The panels were also equipped with detectors that measured the force of the impacts. Three Pegasus flights gave us enough data to indicate that whatever the Baker-Nunn cameras were recording had little to do with the number of particles in space between the earth and the moon. Beyond that, Pegasus recorded no particle with enough force to penetrate the Command Module. We were reasonably confident, then, that meteoroids would not be a problem.

The Lunar Surface Environment

In the early days of the program, there was considerable speculation about the lunar surface. At one of my first meetings with the President's Scientific Advisory Committee, Thomas Gold of Cornell, who was a member of the committee, told me that his analysis of the lunar surface indicated the presence of what he called "fairy castles"—high dunes, perhaps 100 ft. high, created by lunar dust particles adhering to one another because of electrostatic force and having, as a result, virtually no mass. If the LM were to set down on top of such a dune, it would just sink, and there it would be, buried under 100 ft. of moon dust, completely shut off from light and from communication with earth. There would be nothing we could do.

Obviously, if this analysis were correct, we faced a serious problem. Our task, therefore, was to determine how correct it actually was. Charles Townes (of NASA's Science and Technology Advisory Committee), Thomas Gold, and I reviewed all of the measurements that had been made of the lunar surface, including early measurements of the radar reflection from the surface, infrared measurements, and visual measurements. These measurements, as a whole, gave a very inconsistent picture of the surface's characteristics.

At Townes's suggestion, we made detailed radar mappings at several frequencies to measure the reflectance of particular sections of the surface as a function of frequency. These mappings indicated that the dust layer could not be more than a centimeter or two thick, on the average. After studying this new data, I suggested that the lunar surface resembled glacial till, the only difference being the absence of water, which would make it dry glacial till. Although that was something of a wild guess, it turned out to be almost exactly right.

Gold had another concern, however. He argued that, despite our probably being correct about the dusty layer's depth, we would still have the problem of electrostatic adhesion. He pointed out two problems here. The first was that the dust kicked up by the exhaust from the LM's descent engine might cover the windows to such an extent that the astronauts could not see well enough to land. This was a real concern, and we responded to it by attaching probes to the LM's supports which would indicate to the astronauts that the LM was within 5 ft. of the surface and that they should shut down the engine. The second problem was one the astronauts might encounter during the moon walk, when dust kicked up by their feet adhered to their visors and blinded them (see Fig. 4). This possibility, too, was a real concern, and we did try to find a visor material that would not accept

Figure 4 Astronaut working on the moon. (NASA.)

a charge, but we were not able to find one which would resist a charge and, at the same time, satisfy all the other requirements. As a result, we did not know until after the first landing that this concern was unnecessary. (In this instance, as in others, it was difficult to produce on earth the kinds of imaginary conditions we could postulate for the moon—because many of them were, in fact, imaginary.)

The Apollo Flight Design

Our major systems-engineering problem was, of course, designing the lunar flight itself. In the flight design, we had to go deeply into the multitude of tradeoffs that existed regarding the weight of the launch and space vehicles; the thrust various weights would require; the number and location of rendezvous and orbits; the amount of fuel each alternative flight configuration would require (and the weight of that fuel); and, always, the cost. At the outset, one idea was to have a launch vehicle large enough to fly directly to the moon and land another vehicle on the surface without any rendezvous. The vehicle that was being proposed, called the NOVA, would have been about twice the size of a Saturn V, and its design was well along before people realized that in order to build it, we would have to build first a Saturn I, then a larger Saturn, and then the NOVA. The total cost of building these vehicles would be far more than the $20 billion we had budgeted for the entire program. That ended the NOVA, and the no-rendezvous design went with it. Cost notwithstanding, however, there were two other things about the NOVA that raised questions we found we had to consider in selecting a launch vehicle: the amount of noise it would generate and its explosive equivalent. We found that a vehicle the size of the NOVA would generate enough noise at launch to possibly cause serious damage to the area surrounding Cape Kennedy, and that its explosive equivalent as it sat on the launch pad—10 to 15 million lb of dynamite—was formidable: about one-third the force of the Hiroshima atomic bomb.

We had several rendezvous alternatives, each with its group of advocates. One group favored earth-orbit rendezvous; another favored lunar-orbit rendezvous. Another said we ought to rendezvous in earth orbit, assemble a spacecraft for the traverse to the moon, go into lunar orbit, and take a third vehicle to the lunar surface, because that would be the most efficient design, as it would require the least amount of propellants. Still another group said that a double-rendezvous mission would be too complicated—that, instead, we should launch two or three vehicles at once, rendezvous and transfer fuel on the way to the moon, and land directly, without a lunar orbit. This would have meant, of course, carrying enough fuel aboard the landing craft to launch directly back to earth from the lunar surface. This proposal pointed to the basic question: how large should the launch vehicle be, since all of the alternatives being proposed were feasible.

For a variety of reasons, we decided on a single-rendezvous mission. Having decided that, classical mechanics told us that the rendezvous should be made in lunar orbit. It then followed that we wanted to select the smallest launch vehicle capable of the single-rendezvous lunar-orbit mission, which was a Saturn IV (that is, a Saturn vehicle with four J-2 engines). Wernher von Braun, however, pointed out that the Saturn IV design left a space in the center of the engine section that could be filled with a fifth engine (see Fig. 5). Therefore, systems engineering being an exact science, we decided to use Saturn V.

Testing the Saturn V

The first Saturn V launch (V-01) was unmanned, but it was the first time that the first test launch of a vehicle designed for manned flight was made with the vehicle itself in complete mission configuration, including a working Command Module. Many people were worried about the financial risks of such a procedure (I, particularly, since I had argued that it was a good thing to do), but the flight worked perfectly, and it looked as if we were now off to the moon. We had, however, made the rule that we would have to have two successful test flights before we proceeded with the mission, and so we launched V-02. Making two test flights turned out to be a good rule, because V-02 went wrong practically from the beginning. The first stage got out of sight in good order, but when we looked at the records we saw that the POGO[1] had built up to about eight Gs. This was not supposed to happen, and since the vehicle was not designed to take that kind of stress, we were seriously wondering whether it would hold together. It did, probably because it was overdesigned by a factor of 2 or so.

The second stage ignited and was going fine until about 80 seconds into the flight, when one of the engines suddenly shut down and was quickly followed by a second engine. The records indicated that the temperature in the boattail section had gone up drastically, which meant that there had been a bad fire down there. The remaining three engines ran for the rest of their period and shut down on schedule, and the third stage ignited; we felt that although something had happened, at least the craft had not blown up. The third stage burned satisfactorily through its first burn—it was a two-burn operation, duplicating the actual moon flight—and then it shut down on schedule. But when the time came for its second burn, it would not start. Its records showed that it, too, had developed a fire. We had no idea what was wrong.

At the press conference afterward, the press were well aware that we had a problem, because the Command Module had come down several thousand miles off target. They assumed that the flight was a major

[1] POGO refers to the longitudinal vibration of the launch vehicle, which, in this instance, built up to a force eight times that of gravity, much as if the vehicle was a giant pogo stick.

Figure 5 The five engines of the first stage of the Saturn V launch vehicle. (NASA.)

disaster, and asked what we were going to do now. In answering them, I said that I did not know whether it was a disaster or not, that we might have learned enough to feel more confident about going forward with the program. This kind of reasoning was not highly popular, and I had more headlines for that bit of reckless abandon than I had about anything else I have said. I believed then, however, and later events bore me out, that we would learn a great deal from the flight and would proceed with the program.

We did proceed, but the decision to do so was not made until we had gone through one of the most interesting technological detective episodes I have ever witnessed. What had caused the POGO in the first stage was an oscillation that had set up in the oxygen-feed line—a coupled oscillation between the pump-head pressure and the tuned length of the feed line, which was tuned by the level of oxygen in the stage. As the level went down, it caused the stage to pass through the resonant frequency, and the oscillations built up. Fortunately, they died down before the engine broke loose.

After studying the problem for some time, we discovered that blowing a stream of air into the top of the shut-off valves for the stage created a small gas-pressure chamber that damped the oscillations in the feed line. Thus, without any change except a gas line going over to the top of the valve, we were able to cure the problem. In the process, however, we developed one of the most sophisticated mathematical models that exist of the dynamics of hydraulic fuel flow, pump-pressure sensitivity, and several other phenomena. Fortunately, we had a baseline from which to start, because the Titan II launch vehicle had experienced a similar problem. And although the Titan II problem had been solved in a different way, the modeling techniques had been started, so we simply took over the techniques and extended them. One thing we learned—and should have required from the beginning—was that each engine's output thrust should be independent of the pressure of the inlet to the pump which feeds it, in order that the oscillations would not be amplified. (One way to accomplish this was to put in an active feedback circuit so the pump valve will be controlled by the thrust. This approach seemed to be a more straightforward and positive way of solving the problem than trying to tune all the feed lines to keep them from resonating.) Once we eliminated the amplification through the engine, we had no more POGO problems, since the entire launch vehicle was no longer acting like an unstable feedback system.

The problem with the second and third stage engines was altogether different. We had tested these engines on the ground under a great variety of conditions, including vacuum chambers, but we had not been able to duplicate exactly the conditions in space. In particular, we had not been able to achieve a perfect vacuum. As a result, when liquid hydrogen was passed through a set of bellows which led from the bypass line serving as a starter for the engine, the bellows—which were about three inches long and one-half inch in diameter—became coated with ice or frost. The frost damped the oscillations of the bellows. We were never aware of this problem on the ground. In space, however, with no frost forming to damp their oscillations, they broke, because every third or fourth line of bellows was resonant to the flow of liquid through it. When we were finally able to put the bellows in a high vacuum, we duplicated the failure. The most obvious thing to do was get rid of the bellows, which we did. Thus ending the problems with those engines.

Apollo Reliability

Of course, establishing the reliability of the launch and space vehicles was a critical concern—and, not surprisingly, one about which few people agreed. The Saturn V, with the Apollo spacecraft and the launch-support equipment, represents about 15-million parts. If all of those parts must work for the mission to succeed, the reliability of each part must be extremely high. Even a reliability figure of .9999999 for every part does not guarantee a successful mission. Counting piece parts, and using conventional techniques like single-point-failure analyses and critical-path analyses, we calculated about a .5 probability of a successful lunar landing. On that basis, one of every two flights would fail, which was clearly not good enough.

We then went back and started from the beginning and used all of the techniques we had ever been able to discover for improving the reliability of the piece parts. We introduced the zero-defects concept, and we tried to make sure everybody in the program understood how important it was to do everything right the first time. Then, during the tests for the first manned flight on the Saturn IB, the J-2 engine failed the preflight test. When we took it apart, we found a dust cap which someone had left on the engine had prevented the valves from opening. That did not improve our feeling of confidence about the zero-defects program, or about how well the design was actually going to work.

Consequently, in planning the Apollo VIII flight, we added a new technique. We identified the mission's critical events, and asked the two people in the program, who had the most experience with respect to each event, to evaluate the probability of successfully getting through the event. We then built up a probability of .9 for a successful mission and a probability of catastrophic failure of less than .01, which was probably the best measure we ever made of the real chances of success for the mission. In my opinion this analysis, which was purely subjective, but was based on a great deal of experience, was a fair measure of the actual risks involved in the Apollo program; the experience to date, at least, has corroborated it. The moral, I suppose, is human judgment, properly applied, is sometimes better than statistics.

The Apollo Fire

But there are times when neither statistics nor human judgment—nor, indeed, the two combined—are flawless. The disastrous Apollo fire is a stark reminder of our fallibility. We had carefully established what we believed were correct standards for the materials used in the spacecraft. We had run fire tests on the materials in sample-size lots, and we had established what we would do in the event of a fire. We had experienced several short circuits in the Gemini program, where something ignited but was promptly extinguished, so we did feel we understood the problem. But we failed to recognize—and this was a failure in human judgment—the need for a full-scale test on the cabin with all the materials in place in which we simply put a match to it to see what would happen. Had we used such a test, we would have found that in an enclosed volume the size of the Command Module, containing flammable materials, the reflection from the wall would be such that once a fire started the heat would cause the oxygen pressure to build up suddenly, which, in turn, would accelerate the combustion process, cause it to flash, use up all the oxygen, and go out—all within about 60 seconds. That is what happened; in fact, relatively little of the combustible material actually burned.

The personal, political, and philosophical implications of that disaster were grave and affected all of us. Although from a technological point of view we learned a great deal about how to design things to avoid fires, that knowledge was terribly costly. Probably the most fundamental lesson we learned was that ignition sources can never be completely avoided, and the overall design must compensate for that fact. Hence the requirement for thorough testing of full-scale models—a requirement which can be applied to a house or an automobile, as well as to an Apollo Command Module.

The Future

The next major development in the United States space program—in fact, the development on which any future United States space program may depend—is the Space Shuttle, which is based on the concept of reusability. The Shuttle vehicle will be launched vertically (like a rocket), go into orbit and discharge its payload, and return to land at the launch pad, ready to be launched again. Although the Shuttle concept has been in the thinking stages for some time, the technology for accomplishing it has only recently become available. Many people have recognized the value of the concept, but Francis Clauser of Caltech[2] most elegantly put the argument that reusability was a considerable improvement over throwing away $100 million worth of equipment with every launch. He showed conclusively that

[2] F. Clauser, "No Law Says Space Must Be Expensive," *Astronautics and Aeronautics,* **7,** No. 5., (May 1969), 32–38.

reusable vehicles would reduce the costs of space flight by a factor of 100 to 1000. One consequence of this improvement is that the costs will be sufficiently low so that I can look forward to the time when I, myself, can fly in space. I suppose both Clauser and I had this idea in mind when we advanced the reusability idea.

BIBLIOGRAPHY

An Introduction to the Evaluation of Reliability Programs, NASA SP-6501, Office of Technology Utilization, National Aeronautics and Space Administration, 1967.

Bauer, Raymond A., et al., *NASA Planning and Decision Making,* Vols. I and II, Harvard Graduate School of Business Administration, NGR 22-007-163, 1969.

Bedini, Silvio, Wernher von Braun, and Fred L. Whipple, *Moon: Man's Greatest Adventure,* Harry N. Abrams, 1970.

Booker, P. J., G. C. Frewer, and G. K. C. Pardoe, *Project Apollo: The Way to the Moon,* American Elsevier, 1969.

Elements of Design Review for Space Systems, NASA SP-6502, National Aeronautics and Space Administration, 1967.

Introduction to the Derivation of Mission Requirements Profiles for System Elements, NASA SP-6503: National Aeronautics and Space Administration, 1967.

Lay, Beirne, Jr., *Earthbound Astronauts: The Builders of Apollo Saturn,* Prentice-Hall, 1971.

Levy, Lillian, *Space: Its Impact on Man and Society,* Norton, 1965.

Logsdon, John M., *The Decision to Go to the Moon,* The M.I.T. Press, 1971.

Mueller, George E., "Manned Space Flights: Programs, Progress, and Prospects," *Scientific Progress and Human Values,* Edward and Elizabeth Hutchings, (Ed.), American Elsevier, 1967.

Mueller, George E., and E. R. Spangler, *Communication Satellites,* Wiley, 1964.

NASA, PERT, and Companion Cost System Handbook, National Aeronautics and Space Administration, October 30, 1962.

Space Research: Directions for the Future, Publication 1403, National Academy of Sciences-National Research Council, 1966.

The Space Industry: America's Newest Giant, by the Editors of Fortune, Prentice-Hall, 1962.

The Systems Approach to Management (An Annotated Bibliography), NASA SP-7501: National Aeronautics and Space Administration, 1969.

This New Ocean: A History of Project Mercury, NASA SP-4201, Office of Technology Utilization, National Aeronautics and Space Administration, 1966.

Webb, James E., *Space Age Management,* McGraw-Hill, 1969.

"What Made Apollo a Success?" *Astronautics and Aeronautics,* **8,** No. 3, (March 1970), Special Section.

Wilford, John N., *We Reach the Moon,* Bantam, 1969.

9

Planning—Programming—Budgeting Systems

HENRY S. ROWEN

Professor of Public Management
Graduate School of Business
Stanford University

Henry S. Rowen is a Professor of Public Management at Stanford University. From 1967 to 1972 he was the President of The Rand Corporation. He has degrees in chemical engineering and industrial management from M.I.T. and in economics from Oxford University. He has been a member of Rand's Economics Department and the Center for International Affairs at Harvard University. In 1961 he was appointed Deputy Assistant Secretary of Defense for International Security Affairs. From 1965 until he returned to Rand as President in 1967, he was the Assistant Director of the Bureau of the Budget in the Executive Office of the President.

I feel somewhat on the defensive in writing on behalf of a rational method of making decisions in a period when rationality is seemingly out of fashion. On the other hand, I feel strengthened by the knowledge that there really is not a good alternative at the present time to something like the systematic approach to decision making that is my topic.

Planning, programming, budgeting—the words do not sparkle. But if the actions of governments, and for that matter of the private sector, are to be conducted in a way other than accidental, there must be some manner of deciding what needs doing, choosing among courses of actions, and carrying out the decisions made. This, of course, is what the political process is for. But this is equally true in countries that function poorly and countries that function quite well. More needs to be said.

Health Care Legislation as an Example

Let me give you an example. In 1965, after years of effort by various groups, health care legislation was passed by the Congress that came to be called Medicare and Medicaid. These programs were designed to provide health services to the elderly and the poor, respectively. Now what was the nature of the argument offered in support of this legislation? Part of the case dealt with the *health status* of these two groups in our population. The poor had a greater incidence of morbidity and mortality, including strikingly high infant mortality, than did higher-income groups. The health problems of the elderly need no comment. The other part of the argument dealt with the access of these groups to *health services* (which is a very different thing from the status of their health). Access was generally low and also highly variable from community to community and person to person. Moreover, it could be shown that health crises could wipe out the financial resources of members of these groups, and of other groups as well.

But what was the legislation really supposed to accomplish? Here things were vague. Forecasts were made, based on inadequate evidence, of the likely increase in usage of services by these groups and estimates were made of the cost to the federal budget and the dollar savings to the beneficiaries. (Incidentally, I doubt that an estimate was made of the financial benefits to an unintended class of beneficiaries, the nation's physicians.) Some, but too little, attention was paid to the problem of supplying the additional services that would be demanded.

This is what happened. The demand was there, but the supply was not. Physicians' visits per person in our population actually declined from the early to the late 1960s. Hospital days per capita increased at the same rate after the legislation as before. More importantly, little attention had been paid to the reimbursement mechanism for hospitals; hospitals could be reimbursed for their costs no matter how they produced their care. As a result there was an enormous increase in the cost of health services in the years after the legislation was passed. By now the annual cost to the federal budget is more than double what was forecast in the mid-1960s. To be sure, some extra services have gone to the intended groups, judging from data on visits to physicians and length of stay in hospital. But since the total supply increased very little, someone had to lose these services. In the case of physician services, the losers were not the well-to-do but the near-poor, those just above the cutoff line for Medicaid. The question of who paid financially is more complicated. (I should point out that to some of its advocates, this legislation was not intended mainly to improve the health of the beneficiaries but to transfer income to them in a politically feasible way.) But what about the physicians' incomes? Had the

American Medical Association reflected on this outcome, it might not have been so opposed to socialized medicine. Indeed, the AMA is now lobbying for its own version of a National Health Insurance Plan.

The outcome, so far, is not very bad. Or is it? Certainly there were some missed opportunities: opportunities to use this legislation as a way of encouraging the development of new types of health delivery systems, as the Nixon administration is doing; opportunities to try out various ways of financing health care; opportunities to reimburse suppliers of services on a basis other than "reasonable and proper" or "usual and customary." Much could have been done better had there been a period of trial and experimentation before getting committed to a single approach, nationwide. This may seem wildly impractical given the nature of the political pressures and the demand for action. But my point is that unless we think carefully about what we are doing or are about to undertake, we will go from one poorly designed and executed program to another. This caution applies to the reform of our welfare system, the design of our strategic nuclear forces, the setting of autoemission criteria, and many other activities. I am not suggesting that this country is doing nothing on these or other programs. Recently, there has been more analysis on a variety of programs than there used to be, but not enough and not sufficiently connected to governmental actions taken.

Need for Improvements in Government Decision Making

The need for improvements in government decision making can hardly be exaggerated. For example, once a year the Council of the City of Los Angeles considers the budget for the next fiscal year. If this year's budget is like last year's budget, it will be (a) much higher in costs, (b) uninformative about what specific public objectives are to be accomplished by this budget, (c) not cast in such a form that the council members can get a sense of where the benefits and costs are greater or smaller at the margin. In these respects the legislature of Los Angeles is no worse off, and is perhaps somewhat better off, than comparable bodies in other cities. Nor is this a problem of legislative bodies alone. Often the executive agencies are not much better informed. They know a great deal about their current activities, but the relationship between their activities and activities of other agencies of government or the wider public interest either may not receive much scrutiny or may be dominated by the parochial goals of the agencies.

Obstacles to Systematic Decision Making

There are, however, several obstacles to systematic decision making in the public sector.

One problem is that there are always hostages to the past. The discretionary part of an agency's budget in any given year is small because prior year decisions have committed most of the resources currently available. Moreover, although civil servants often take the long view, politicians usually do not. They characteristically have a high rate of time discount. This propensity of politicians taken together with the small portion of the budget that is discretionary at any time means that programs frequently get started without a far enough look ahead to see what is needed to finish them and without a meaningful commitment to completion.

The temptation to "make a splash" in announcing new programs is often very great. Besides, the sponsor probably will not be around to see how the programs come out anyway. I am not saying that this is done deliberately; more often, it is inadvertent. This propensity is enhanced by the mismatch between the time it takes to plan and execute a major public program and the frequency of our election cycle. The program may take a decade or more to plan, test, implement and "de-bug;" but, elections seem always with us. We need to have the opportunity to change administrations regularly, but the price we pay for frequent elections is high. This price is the large and seriously underestimated cost in work begun and not finished and in the foreshortened time perspective such an arrangement produces.

This situation is closely related to the problem of cutting back on obsolete programs. These are programs that might have made sense when they were started, but long since have lost their usefulness and refuse to go away. Some agricultural support programs are classic examples. The estimated present value of annual subsidies in a number of these programs is over $100,000 per recipient. One could make handsome gifts to people in certain occupations, if they would promise to work in another field. Our maritime subsidy programs provide other splendid examples.

A third difficulty is that many public issues and programs are very hard to evaluate. To be sure, suboptimization can often carry us quite far without too great a sense of discomfort. At the other end of the scale, the highest level issues that confront our society—those frequently discussed under the heading of national goals—are quite beyond our formal analytic powers. It is in the important middle ground, that occupied by Head Start programs, job training programs, the space shuttle, and the like, where one can legitimately aspire to have important meaningful analysis on important issues. But, not with ease. Our theories and measurement techniques are often inadequate.

Fourth, frequently the responsible agencies do not have much incentive to apply analysis techniques even when they are available. For one thing the results might be unfavorable to that agency or its program.

Fifth, there is the complicated business of estimating the costs and benefits not just in the aggregate, but to different groups and individuals. For it is not good enough to estimate these aggregates—and that may be hard enough—but one must understand *who* gains and loses. This is so partly because national goals are not givens; they emerge from the workings of the political process, and also because ends and means, in general, are not independent. *How* they are done can be as important as the objective to be served.

Finally, there is the failure to hold managers to their promises. All too often, when actual performance falls short of prediction, little notice is taken. And, of course, one way of protecting oneself against this risk is not to make specific promises or forecasts in the first place—which is a good reason to insist on them.

With these six barriers to systematic decision making, it is not surprising that we do poorly.

Four years ago, I wrote the following words which seem to still apply:

> In all of these areas the problem is not so much lack of knowledge, as lack of knowledge relevant to policy; not so much lack of technique, as lack of technique for analyzing policy questions; and not so much lack of skills, as lack of skilled people working together on these problems. What must be found is some way to harness our knowledge of social processes, to sharpen our process of analysis, and to involve and train good people in methods of structuring and solving complex policy questions.
>
> These problem areas are, in large part, the concerns of government at all levels. But, they lack the means for tackling these problems effectively. At the federal level few departments have an analytic capacity of any consequence. They generally lack the ability to identify problems with precision, see that relevant data are collected, policy alternatives are formulated, valid experiments are run, and existing programs are evaluated. Even the simplest fact, such as finding out how many able-bodied males in this country are supported by welfare, has taken many months for a high official to acquire. Straightfor-

ward data on health care, education, ghetto jobs, and housing are often inadequate. The more complex question, 'What is really being accomplished by these programs?' is rarely addressed in depth.

A Better Approach

Planning, programming, and budgeting procedures are one approach to trying to do better. It has a multiplicity of origins. From engineering, through the design of complex systems; from economics, through the development of cost-benefit techniques of analysis; from public administration and business, through the development of program budgeting. It received a considerable impetus in the defense department beginning in the early 1960s and was adopted generally by the Johnson administration in 1965.

Its essential aspects are:

- A careful specification and a systematic analysis of objectives.
- A search for the relevant alternatives, the different ways of achieving objectives.
- An estimate of the total costs of each alternative—both direct and indirect costs, both initial costs and those to which the alternative commits us for future years, both dollar costs and those costs that cannot be measured in dollar terms.
- An estimate of the effectiveness of each alternative, of how close it comes to satisfying various objectives.
- A comparison and analysis of the alternatives, seeking that combination of alternatives that promises the greatest effectiveness, for given resources, in achieving the objectives.

The Component Parts of PPBS

Technically, the component parts of the federal government's PPBS, in the latest version with which I am familiar, are: (1) *program structures,* which display each agency's *physical and financial activities* according to objective or common outputs; (2) *issue letters,* which summarize the agency's and the Office of Management and Budget's list of major policy issues in need of analysis and evaluation during each planning and budgeting cycle, and *special analytic studies,* which reflect intensive analysis of these issues; (3) *program memoranda,* which register agency choices between alternatives and summarize relevant analysis affecting the decisions; (4) *program and financial plans,* which display for the past two and for the next five years data on the financial inputs and physical outputs resulting from proposed and past commitments.

How Has It Worked?

How, then, has it worked? On the whole, not well. The obstacles to satisfactory decision making I described earlier have impeded the successful use of PPBS.

For example, there is a problem in getting agencies to present serious program and financial plans. If the President asks them to prepare plans without giving them a financial constraint, what he gets back are lists of what they would like to do, which always cost far more than the money likely to be available. This is not very helpful. If he gives them dollar constraints that reflect a more feasible dollar level, this may cause trouble both with the agencies and with their supporters in Congress. Thus he may not want to take on what might seem to him like an unnecessary fight.

There has been more success in getting analyses of issues carried out. Some excellent analytic work does get done. But often the results are quite unsatisfactory, sometimes because of a shortage of time or competent people; often because the state of knowledge is simply inadequate to support more than the most superficial kind of analysis. One problem, which I might have listed earlier as a general obstacle, is that many domestic programs are financed in large part by federal agencies, but carried out by local governments. The federal agencies, not being responsible for operations, do not have direct access to operational data on these programs; nor do they necessarily have a strong incentive to find out what works and what does not. The local agencies also have a weakened incentive to find out what works because much, perhaps most, of the costs are being paid by the federal government. There is an argument here for revenue sharing, *if* the governments on the receiving end come to regard the shared revenue as really being theirs.

For these reasons the Nixon administration probably will not continue the PPBS system in its present form. Nor should it. It very likely should and will give increased emphasis to the *analysis of issues.* And it will retain some kind of a *program budget,* one, for example, which could show how much health care is provided in the aggregate from all the federal agencies. However, it probably will not continue to demand the development of program and financial plans or routine program memoranda.

Criticism of the Process

As I said at the outset, this subject is not very fashionable today. There was an interesting article in the *Village Voice* recently about the work of the New York City-Rand Institute. This is the institute that Rand has set up jointly with the City of New York to work on that city's problems. It was not a friendly piece. Much of the criticism was to the effect that our work with the city was concerned with questions of *efficiency* in the delivery of the city's services, such as in housing, fire, health care, and corrections. The writer claimed this was wrong, because the whole system was bad and had to be changed. On the other hand, he did not have a proposal on how to change it or what to change it to. I have a modicum of sympathy with his view. It is clear that fundamental problems do need more attention. However, it is also important to make things function tolerably well. And often they do not.

The writer also was critical of the fact that much of the work being done was not being exposed enough to public review and criticism. There, he was on sounder ground. A necessary element in getting better public policies adopted is through a process of open review and criticism. But government agencies, at all levels, often resist this.

This criticism bears on the nature of the political process and the role of policy analysis in this process. The best descriptive model we have of the public policy process is the partisan, mutual bargaining model of C. E. Lindblom.[1] He describes the process by which public decisions are made in our society as one in which different groups, each with its own objectives, enter the political market-place to achieve certain purposes. Through the interaction of these groups, with overlapping or conflicting values, compromises are struck and decisions emerge. Central to this view is the point I made earlier that means cannot be clearly separated from ends. They interact strongly and, by and large, people do not have preferences for ends, which are often abstract, but they do for means, which are concrete and "real." This is an excellent description of how things work, and it has powerful normative implications as well.

Systematic Analysis and the Political Process

How can systematic analysis be fitted into this view of the political process? I have long advocated the creation of a multiplicity of analytic groups within executive branches of government, in legislative bodies and outside of government altogether, each able to address many of the major public policy issues facing the country from differing perspectives. This is critically important because only through a

[1] C. E. Lindblom, *The Intelligence of Democracy: Decision Making Through Mutual Adjustment,* Free Press of Glencoe, 1965.

process of review and criticism can inadequacies and biases be exposed, alternative hypotheses raised, and additional data surfaced. This is, I believe, a necessary condition (but not a sufficient one) for an improved public policy process. It recognizes that even the best informed and intentioned analysts are likely to be affected by their environment. They will, perhaps unconsciously, formulate alternatives and choose evidence in ways most comfortable within their organizational setting. The best way to deal with this problem is through a structured adversary process.

This proposal suggests a role for analysis for Congress and other legislative bodies. There has been a growing interest within Congress in improving its own analytic capacity. Hearings have been held on this subject and a variety of proposals have been put forward. They include building up the capacity of the General Accounting Office or the Library of Congress, creating a new research institute for Congress, or increasing the access of committee staffs to outside analyses. However it is done, there is a good case for strengthening the analytic resources of the Congress. I note, by the way, that the California legislature has pioneered here with the creation years ago of the position of Legislative Analyst, a most useful function.

The Future

Finally, where should we go from here in the applications of systematic analysis to public policy issues?

In addition to having more participants in the analytic process we simply need to do much more analysis. Many important questions continue to be under-investigated.

Second, much more emphasis needs to be placed on getting empirical data. Whether in weapons development, the choice of a health services system, or alternatives to the present welfare system, many more experiments and operational tests are needed before commitment to major programs.

Next, the scope of analyses needs to be widened. Murray Gell-Mann, Professor of Physics at Caltech, says we must find ways of incorporating intangibles such as beauty, effects on the environment, and other values more directly into our analyses. I think that much can be done here. Clearly we should not try to reduce all of our values to a single parameter, as is done in studies in which both the benefits and the costs are expressed in dollar terms. In these analyses, if one has a benefit-cost ratio greater than unity (using the appropriate discount rate—itself a sticky problem), then the project should be carried out, and if not, not. This approach, used often in public works projects, sometimes includes so-called "secondary" benefits (those less tangible but for which numbers are devised), often in order to shove the benefit-cost ratio over unity so that the project will get a go-ahead decision. Although measurement of costs and benefits in dollar terms has its use, in general, the objective function should not be expressed in single-value terms. Nor are all of the relevant costs necessarily dollar ones. Usually one has a variety of objectives, and the alternatives under consideration offer different combinations of payoffs. Then the choice can only be made subjectively, but nonetheless, more systematically.

Another approach is to do more to establish rank order preferences among objectives, possibly through greater use of polling techniques and through simulation techniques which seek to elicit people's preferences. I am not advocating naive attempts to elicit preferences from people on matters about which they are not clear and for which, in a real sense, they do not know their preferences. We just do not carry around in our heads comprehensive, well-mapped value preferences over all of the choices we might face. But, there are ways of exposing people to many of the relevant choices and then getting some useful feedback from them.

Still another is to pay more attention to organizational incentives. Organizations often have deeply imbedded missions and the people in them behave in ways that conform to their character. This is a powerful force and one that has been neglected in the tradition of systems analysis. It has not been neglected by everyone and the Nixon administration is devoting a good deal of energy to this problem—as is evidenced by their proposals to change the structure of government.

Finally, I would emphasize the utility of regarding this entire subject in terms of invention, that is, as a design problem. Many of these problems appear, and are, intractable if viewed statically. The choice among incommensurable alternatives in a partisan, mutual adjustment process is enough to make one throw up one's hands—as some people have. But new inventions, both technical and social, can make a difference. They introduce a new set of alternatives which may help—not necessarily by providing complete solutions, for new inventions create new problems, but by ameliorating and resolving old ones.

In sum, I have tried to give a reasonably balanced view. How you regard it depends on your biases. If you are impressed by the power of systematic analysis, you may be depressed by this exposition. If you are of the view that the analysts have had too much influence, take heart, their influence is pretty limited. Whatever your views, the subject is a fundamental one and there is no doubt that there is plenty of room for improvement both in understanding and in practice.

BIBLIOGRAPHY

A Bibliography of Selected Rand Publications: Systems Analysis, The Rand Corporation, SB-1022, 1967.

Bauer, Raymond A., and Kenneth J. Gergen, *The Study of Policy Formation,* MacMillan, 1968.

Braybrooke, David, and Charles E. Lindblom, *A Strategy of Decision,* Free Press of Glencoe, 1963.

Hinrichs, H. H., and G. M. Taylor, *Program Budgeting and Benefit-Cost Analysis,* Goodyear, 1969.

Hitch, Charles J., and Roland N. McKean, *The Economics of Defense in the Nuclear Age,* Harvard University Press, 1961.

Lindblom, Charles E., *The Intelligence of Democracy: Decision Making Through Mutual Adjustment,* Free Press of Glencoe, 1965.

Hovey, Harold A., *The Planning-Programming-Budgeting Approach to Government Decision Making,* Praeger, 1968.

Lyden, F. J., and E. G. Miller, (Ed.), *Planning Programming Budgeting: A Systems Approach to Management,* Markham, 1967.

McKean, Ronald N., *Efficiency in Government Through Systems Analysis,* Wiley, 1958.

McKean, Roland N., *Public Spending,* McGraw-Hill, 1968.

National Goals Research Staff, *Toward Balanced Growth: Quantity with Quality,* U.S. Government Printing Office, 1970.

Novick, David, *Program Budgeting: Program Analysis and the Federal Budget,* Harvard University Press, 1967.

Rivlin, Alice, *Systematic Thinking for Social Action,* The Brookings Institution, 1971.

Schultze, Charles L., *Setting National Priorities: The 1971 Budget,* The Brookings Institution, 1970.

The Analysis and Evaluation of Public Expenditures: The PPB System, Vols. I, II, and III, U.S. Government Printing Office, 1969.

10

Systems Concepts in Social Systems

ROBERT BOGUSLAW

Professor of Sociology
Department of Sociology
Washington University

Robert Boguslaw is a Professor of Sociology and formerly was chairman of the Department of Sociology at Washington University in St. Louis, Missouri. He has engaged in research at The Rand Corporation and the System Development Corporation on the characteristics of man-machine systems and has participated in interdisciplinary research on large-scale information systems and social systems. He has conducted research in the areas of national policy formation and social planning and has been affiliated with the Institute of Labor Relations and the Research Center for Human Relations. He is the author of the C. Wright Mills Award Winning book, *The New Utopians*.

I should like to make some observations on the current state of the profession of engineering and relate these observations to some problems involved in the design of social systems. I do not believe it is meaningful to discuss one of these issues without the other.

The Profession of Engineering

At this particular time in American history, it seems clear that the profession of engineering in this country is facing a rather remarkable change in the conditions of its own environment. For the first time in the lives of most engineers, and certainly for the first time in the lives of current engineering students, the prospect arises that something like widespread unemployment may exist even for well-trained, highly competent professionals.

Now, the duration of this apparent crisis, and even the question of whether the crisis is "real" or not, are not matters I wish to consider here. I am not at the moment concerned with the economic issue of the nature of employment curves in this profession. I do wish to suggest, however, that the *prospect* of unemployment or the *belief* in its possibility, no less than the actual *fact* of unemployment, may well have salutary effects on the attitudes, values, and behavior of persons in this profession. And these, in turn, may well have important implications for the conceptualization of social system design.

For example, as long as the bubble of prosperity is expanding, and upward mobility is a well established characteristic of engineering careers, the more ambitious professionals tend to cast their eyes upward to see what lies at the top of the career ladder. If they allow their eyes to move high enough, they ultimately see something called "management"—because that is the name mid-century America gave to the apex of the career pyramid. But when the bubble breaks, managers at all levels tend to join the ranks of the unemployed even more rapidly than do members of the hard core technical staff. And under these conditions, for many working engineering managers, no less than for students and established technical professionals, there arises the possibility that one may be required to engage in technical work for the remainder of his career—if, indeed, jobs continue to be at all available.

But this very possibility carries with it the potential for drastic changes in the value structure and attitudes of members of the profession. If you no longer identify with management, with what can you identify? On the new career ladder, which way is "up"? Thoughts such as the following arise: How realistic is it to expect that existing places of employment will address themselves to new tasks for which society is willing to pay? If it develops that reconversion expectations are not realistic, do there exist alternative settings or can alternative settings be created within which engineers can use their talents and knowledge? If I can no longer practice my profession within the cloistered walls of an aerospace company, what on earth do I do?

To raise such questions is implicitly to raise questions about both the nature of the engineering profession itself and the characteristics of the layer of society within which it finds itself. In the course of daily study and work one gets a somewhat distorted notion of any occupation. What, in point of fact, is the basic character of engineering as a profession?

For many it seems to come down to a familiarity with the characteristics of certain specialized materials and some information about how these materials can be redistributed in certain more or less definable ways through the use of some well-established methods or "techniques." The dictionary will tell you that engineering is a discipline that concerns itself with the properties of matter and the sources of energy in nature and tries to make them useful to man.

Within the framework of this definition human beings find themselves in an anomalous position. On one hand, they are the "man" for whom all this activity is made useful; on the other hand, they become one of the energy sources in nature that may itself be utilized in a transformation process.

Everyone recalls tales of how the pyramids in ancient Egypt were assembled almost exclusively through the use of human energy. It is fashionable to believe that, in a highly industrialized society like twentieth century America, man is no longer required as an energy source. This, of course, is simply not true. In point of fact, seeking employment as an engineer or other professional is, in a very real sense, to offer for sale reservoirs of energy for use in certain decision-making, transportation, and communication activities. Despite the phenomenal progress made in the field of computerized systems, it is probably safe to say that the truly man-less system is a distinct rarity on the contemporary scene.

The Definition of a Social System

If, then, we think of social systems as systems which embody groups of men (or women) as functioning parts, we in fact are talking about an enormous range of systems with which even "nonsocial-system" engineers are intimately involved. And, alternatively, if we discuss social systems as systems designed to serve some human or social purpose, then by definition we are discussing literally all engineering systems.

In short, if the defining characteristic of social systems were simply the fact that these systems use human components, then we are certainly in a familiar domain for the most conventional of engineers. If we should extend this concept to include systems containing nonhuman components for which human components could readily be substituted, then we are discussing a much larger set. If we reverse the procedure, and examine systems currently using human components which could readily be replaced by computer-guided nonhuman components, then our defining concepts become muddy indeed. We appear to be discussing not conceptually distinct phenomena when we talk about social systems versus nonsocial systems, but rather systems which at a given point in *time* either use or do not use human components.

The characteristics of human components have been thoroughly analyzed, and the varieties of transformation which the human being, considered as a system component, performs upon inputs to produce outputs are called human functions.[1] Thus in the first place there is a *sensing* function which indicates the presence or absence of a difference in physical energies. Second, there is an *identification* function which involves distinguishing among different input qualities for the purpose of making different responses to different classes of inputs. A third function is called *interpreting.* This involves categorizing inputs according to their effects rather than their appearances.

Each of these functions operates under a variety of conditions that have been analyzed to provide guides for selection and tooling or training of the human components for the purpose of achieving the necessary degree of performance, reliability, accuracy, and durability.

Some writers, when discussing configurations composed predominantly of human components, refer to them as social systems, defined as "a set of patterned relations among structured elements so that changes in one element set up pressures for adjustment or other types of changes in remaining units."[2] Other writers tell us that social systems, like other kinds of systems, have a unity which is an "outgrowth of differentiation of parts, interdependence between the parts, and predictability in the connections between the parts."[3]

But of course the notions of differentiation, interdependence, and predictability in connections between parts are familiar notions in the study of predominantly nonsocial systems as well. And many sociologists avoid the use of the term "social system" precisely because it seems to suggest that everything is working smoothly; that the gears of society are meshing well; that the old social boxcar will continue to roll indefinitely, if only people do not tinker with its wheels. Use of the term "social system" also seems to convey the notion that everything in the system has some "function", however remote and indirect it may be. But, of course, this is not at all the case. Many features of social systems represent outmoded adjustments to a previous set of conditions. Other features are simply things which some individuals, groups, or classes of individuals have managed to introduce into the scheme of things because it suits their own private needs for material

[1] Cf. Robert M. Gagné, "Human Functions in Systems," in Robert M. Gagné (Ed.), *Psychological Principles in Systems Development,* Holt, Rinehart and Winston, (1962), 35–75.

[2] Neil J. Smelser, "Introduction," in Neil J. Smelser (Ed.), *Sociology,* Wiley, (1967), 6.

[3] Everett K. Wilson, *Sociology: Rules, Roles and Relationships,* Dorsey Press, (1966), 48.

goods or social power—not because society needs these things in order to survive. But the use of the term social system seems to suggest that it is *really* necessary for crime, wars, and human degradation to exist, since they are found in ongoing social systems.

Moreover, the systems concept seems to suggest that societal effectiveness would somehow be increased if only things worked "more smoothly," that is, if frictions and even differences in opinion were eliminated. In biological organisms we tend to applaud variation. "Vive la difference" is not simply a mating call for sex-hungry adolescents; it embodies a philosophy and a science of growth which asserts that biological variability is a necessary precondition for desirable evolution and ultimately for increased human creativity. But designers accustomed to working with predominantly non-social hardware systems have learned to worry about variability; they equate *variability* in structure and performance with *unreliability*. They therefore see variable systems as poorly designed unreliable systems and their designers as muddle-headed blunderers or worse.

The Difference Between Social and Nonsocial Systems

I think this points up what for me is the core difference between social systems and nonsocial systems. The difference for me is *not* to be found in the relative proportion of human components used in a system complex at any given point in time. Furthermore, the question is not one of reliability versus nonreliability. (Human components can be remarkably and even fantastically reliable in the sense of being inflexible or invariant in the presence of changing conditions. In fact, I personally am convinced that many of the central problems of Western civilization can be traced precisely to the fact that existing human populations tend to be excessively invariant in their filtering devices and internal data processing mechanisms.) The question ultimately is one of the definition of system objectives and, in a much deeper sense, a question of purpose, goals, and values. It depends on what it is that you regard as being important.

If you begin, or are predominantly preoccupied, with materials and how these can be assembled for some given system purpose, then no matter how high the ratio of human components used in your system, you will, in my view, continue to be a designer of nonsocial systems. If, on the other hand, you are predominantly concerned with people and view them *sui generis*—as things important in themselves—if you regard them as more or less self-generating, independent entities which represent the ultimate definers of all system goals—if you are ethnocentric about the race of mankind—then you are somewhat prepared to enter into the activity of social system design.

Who is the Systems Designer?

Let us try to trace some of the implications of these differences. Nonsocial systems characteristically have a system objective formulated by and relevant to the designer or some specified others who stand outside the system itself. Thus it may be decided to design an airplane that will fly at an altitude of X thousand ft above sea level, at a speed of Y mph, carry a payload of Z tons, and so on. These will become the formal system objectives. Convert the same system into a social system. The *elements* or *components* of the system may now wish to define system objectives in terms of their own special values, needs, or requirements. We suddenly seem to be involved in a problem of designing an aircraft because a set of wheels wishes to be off the ground—or, even more perversely, when the wheels really want to stay *on* the ground!

But, says the nonsocial system aircraft designer, "Are you kidding?" "Do you think I'm going to sit here and let a pair of *wheels* tell me how to design my ship?" And the answer, regrettably, is a yes, somewhat. The point is, of course, that if it is to be a social system, it cannot be exclusively a "my" ship. Under many circumstances, it may well have to be a system truly designed according to the specifications of what may eventually be, or even immediately are, its component parts. The perspectives of the designer and components may differ radically. And there must exist some means for adjusting them to each other.

If you are designing an educational system, you had better be ready to deal with something more than the statement of objectives provided by the superintendent of schools, the board of education, the mayor, or even the governor. An educational system, as Californians have good reason to know, is an exceedingly complex affair. Students often have objectives that differ markedly from those of the faculty; the faculty often has objectives which differ significantly from those of the administrators; administrators' objectives may differ drastically from those of various members of the community, and so on.

Under these conditions it may be possible to define one task of the systems designer as essentially similar to the task of a labor arbitrator in industry—that of working out acceptable tradeoffs and compromises; of dealing with the realities and prospects of power; and defining objectives acceptable, if not completely satisfactory, to all parties involved. But this task, of course, is much more complex than that normally assigned to a labor arbitrator. The systems designer may never be aware of the range of people whose objectives are relevant to the design decisions that must be made; these people may not have a formal organization to speak for them, yet one may be developed after the system impact has become manifest. It may become necessary to develop means for predicting and assessing, on an ongoing basis, the impact of a given system on unsuspecting publics of all sorts. The use of simulation laboratories for this purpose has proved to be very helpful, but the development of a useful social systems simulation involves a range of considerations not normally present in the more familiar windtunnel-like techniques used in non-social systems design. Frequently, the objectives of social systems simulation are defined as *searches* for possible relevant analytic dimensions, rather than tests of the impact of previously specified variables. And here I should like to emphasize one critical point: the techniques useful in designing and analyzing predominantly nonsocial systems can almost never be carried over in an uncritical fashion to social systems work. There are countless lurid examples of the consequences of this kind of practice.

Examples of Social versus Nonsocial Analyses

The Value of Human Life

In applying planning-programming-budgeting techniques,[4] one might decide to investigate the motor vehicle accident problem as if one were a rational central decision maker for a national government, whose concern it is to maximize net taxes over a given period of years. From this vantage point, it is possible to analyze the short-run or long-term tax costs of various measures designed to reduce accidents. Within the framework of such an analysis, sooner or later it would become apparent that some accident prevention measures are not justified in terms of the tax benefits derivable from them. In conducting the analysis, it would be necessary and reasonable to place a monetary value on human life, as civil courts frequently do for other purposes.

[4] Cf. Robert Boguslaw, "The Design Perspective in Sociology," in Wendell Bell and James A. Mau, *The Sociology of the Future,* Russell Sage Foundation, (1971), 240–258.

Indeed, to exclude the fiscal value of human life, especially as related to potential tax revenue, would introduce serious errors into the analysis. But the value of human life is not a constant except within a very circumscribed frame of reference. You may place little or no value on my life (especially if I happen to be someone who pays little or no taxes). But to my mother, I am simply priceless. And I may be equally priceless to myself (although the operational definition of priceless may vary considerably with different individuals). If my skin happens to be, let us say black, I may not be worth very much in your analysis, simply because you have not given me an opportunity to participate fully in the economic process. Or you may have denied me the opportunity to obtain the education and training necessary to produce large income taxes for you. And so, for the price of a slight change in your irrational feelings about my skin color, it may be possible to increase greatly the value of my life as carried on your books. Or, for a relatively small outlay of tuition, you may be able to increase my earning potential enormously.

But, if you continue to think along these lines, I will continue to resent you enormously. By what right do you dare imply that saving my life is simply not worth, say, reducing the standard of living slightly for X other people or decreasing their convenience factor by Y? Or, perhaps more accurately, by what scientific or other authority do you assign *any* value to my life in terms of comparative costs and benefits? I have already explained to you that I am really priceless.

Reduction of Welfare and Unemployment Costs

In a similar vein, one could dramatically demonstrate the consequences of inadequate welfare and employment programs in our major cities by calculating the total cost of urban riots in terms of insurance claims, loss of business, property damage, and the like. Such an analysis, if carried out in a rigorous fashion, might well lead to recommendations for preventive measures, such as shooting certain categories of citizen on sight as the most economical means for dealing with the riot problem. One simply locates high risk factors such as hair length, state of clothing repair, or precise shade of skin color, and goes "boom-boom."

Crime Reduction

Several years ago a "science and technology" task force report to a presidential commission on law enforcement and administration of justice made a series of recommendations relating to crime reduction.[5] The recommendations covered a wide range of proposals for research and development, including suggestions for some positively mouth-watering technological innovations: a semiautomatic fingerprint recognition system to replace the existing manual system, new police alarm systems, nonlethal weapons, a national information system for criminal justice agencies, and many others.

In reviewing this report (which was prepared by highly talented engineers and scientists), I was appalled at the narrow definition of the problem adopted. I asked:

> What indeed *is* the system to which all this 'science and technology' is to be applied? At the very least the system must include, not only the cops and robbers, but the larger populations from which these elements have been isolated. One could generate another report addressed to the problem of maintaining surveillance and control of police activities to insure that infringement of civil liberties does not occur. What automatic alarm systems could be devised for individual citizens to protect them against unfair treatment by the police? How about an information processing center and control system to reduce time delays in giving aid under these circumstances? Can one really deal with the problem of Index crimes in isolation from problems of public disorder and 'white collar' crimes? Can one seriously consider proposals for new surveillance equipment, alarm equipment, and control procedures apart from considerations of the social milieu in which they are to be employed?[6]

The point is a simple one: the client of this task force was not really the President of the United States serving as something of an embodiment of the interests of *all* the people. The client of the task force was, in fact, the Department of Justice charged with a set of specific responsibilities relating to crime control. These responsibilities encourage a view of the world from a specialized perspective, which focuses attention on selected components of a more encompassing social system. In this sense the technical staff of the task force becomes identified with the system objectives of professional law enforcement officials. These objectives are circumscribed by assigned duties, resources, and very real problems of law enforcement. It is not within the province of law enforcement officials to concern themselves with nonpolice solutions to the problem of crime, to focus attention on equipment devel-

[5] *Task Force Report: Science and Technology,* Report to the President's Commission on Law Enforcement and Administration of Justice, prepared by the Institute of Defense Analyses, U.S. Government Printing Office, 1967.

[6] Robert Boguslaw, "Review of Task Force Report: Science and Technology," *Law and Society Review,* **II,** No. 2, (February 1968), 287–290.

opment designed to protect the rights of suspects, or to deal with crimes against natural, moral or even prospective law. Yet, all of these may well be implicated in their systems analyses and recommended solutions.

Systems Analysis of Warfare

As my final example let me recall for you an application of game theory used in connection with the systems analysis of warfare.

Thermonuclear war is a social alternative. It is scarcely a new one. For years "defense intellectuals," military men, politicians, and many others have warned us that we must have the courage to do some "hard" thinking about it. In Herman Kahn's classic phrase, we must engage in "thinking about the unthinkable." For, as he puts it, "even if one were to consider thermonuclear war unthinkable, that would not make it impossible."[7]

The implication, of course, is clear. We must free ourselves from Freudian or other resistances to thinking about unpleasant topics and consider the alternatives "rationally." This, in general, is probably a healthy approach to most human problems. The scope of possible outcomes to the thought process, however, is theoretically defined by the range of assumptions used in the analysis.

What "system" is involved in the alternative of thermonuclear war and what are its characteristics?

A fundamental assumption underlying the analysis of thermonuclear war strategy, along with the analysis of many other warfare alternatives, may be stated in the following terms: " . . . national security must be sought primarily through the maximization of national power within international conflict situations. The quest for security is therefore basically a zero-sum game in which each nation's objective is the improvement of its relative power position."[8]

Within the framework of such an assumption, it seems very reasonable to expect that efforts to solve the problem of national security would encompass activities like mobilizing national power resources in conflict situations. The system with which we become concerned initially is something called the nation-state. Implicit in this system definition is an acceptance of force as a legitimate and necessary part of the security-search process.

This, however, violates some deeply held notions about the proper way for nice people to behave. As Kahn and others have observed,

[7] Herman Kahn, *Thinking About the Unthinkable,* Horizon Press, (1962), 39.
[8] Morton Berkowitz and P. G. Bock, (Eds.), *American National Security: A Reader in Theory and Policy,* Free Press, (1965), 63.

except for cases of "just" revolutions (presumably the original American Revolution), Americans prefer to believe that the initiators of force are either sick, criminal, or insane. Such persons should, according to this popular perspective, be categorized as outlaws and either exterminated or treated medically.

But Kahn tells us that "The common American attitude toward force is somewhat naive. Force is a permanent element in human society, used by good, bad, and indifferent nations and people. . . . Even if we unreasonably or immorally institute the use of force, coercion, violence, or threats, it is entirely possible to go on to use these things in a reasonable fashion."[9]

In short, it seems there are two unfortunate traditional biases held by the American people: (1) an unwillingness to initiate the use of force even on a moderate level and for limited objectives, and (2) an undue readiness to use force extravagantly and in an uncontrolled fashion after the initial commitment to use it has been made.[10]

Obviously what we need is an antidote for these proclivities. Kahn provides one by telling us how to escalate force with sophistication. The context within which his analysis proceeds is called an "escalation ladder"—one for the United States and one for the Soviet Union.

But these ladders constitute a retreat from the fundamental assumption about national security, namely, that national security must be sought primarily through maximizing national power in international conflict situations. If we take this assumption at face value, the notion of using an escalation ladder is utterly illogical, if not downright traitorous. Simple common sense seems to say, "Step up to the topmost rung immediately and insure victory." Maximum power is to be found at the topmost rung, and it is precisely this argument that is used by "patriots" who have fears about the national potency image. Of course, within the theoretical rubric of something like Thomas C. Schelling's "strategy of conflict,"[11] immediate escalation to the topmost rung would be seen as rational only in a case of "pure conflict," that is, a set of conditions in which the interests of two conflicting parties are completely opposed.

The strategy of conflict rubric assumes, of course, that this is seldom, if ever, the situation confronting nations in international affairs. With respect to thermonuclear war, I might add it is almost inconceivable that a pure conflict situation could arise. In any event, under the maximization of national power assumption, the search for national

[9] Herman Kahn, *On Escalation! Metaphors and Scenarios,* Praeger, (1965), 17.
[10] *Ibid*
[11] Thomas Schelling, *The Strategy of Conflict,* Harvard University Press, 1960.

security is framed in absolute terms: maximize national power. The escalation ladder is based upon a much more relativistic assumption. It might be expressed as: sufficient unto the rung under discussion is the necessary power thereof.

Instead of an analytic system consisting of the United States as a nation-state, the escalation ladder system embraces the most probable enemy of the United States in a thermonuclear war. And it is not simply the absolute power available to an enemy that is of interest, but rather the enemy's readiness to employ given fractions of its power under varying sets of conditions. Within the framework of the first assumption, the task for the analyst is simply to find ways of increasing the scope of national power. The engineering directive might be stated simply as: "Give me gadgets that will produce the biggest bang for the buck."

Within the framework of the second assumption, however, it is literally impossible to engineer anything without detailed information or speculation about (1) the system of rationality most likely to be used by a potential enemy, (2) the state, not merely of his strength, but of his "resolve," (3) his "value" structure, all supplemented by a wide variety of cultural insights.

All of this serves to make the system of analytic concern enormously more extensive and in many ways less definitive. Special aspects of the analysis may require that previously undreamed of factors be brought into consideration. This has all sorts of important implications for the analysis which I do not want to go into here. The significant point for our present purpose is that a shift in system definition has occurred. Whatever the merits of the escalation ladder concept, its utilization automatically implicates a *different* system than the one with which the analysis began. And it now involves a view of the system from the vantage point of one of the subsystems or components—and again presents to us the mad, but completely authentic, vision we saw before—the vision of an aircraft being designed by a pair of its own wheels.

The System and the Environment

By now it should have become very apparent that specifying what is "system" and what is "environment" has enormous implications for the work of designing systems of all sorts. The environment is simply that part of our cosmos which at any particular time we regard as not subject to change. Changes occur within the system. But throughout human history the question of what is environment has undergone continual modification. When a significant redefinition of system and environment occurs, we are in the presence of a "breakthrough" or "revolution"—scientific, technological, or social. In the logic of tabloid rationality, a scientific or technological revolution is good, while the other kind, in general, is "bad." A technological or scientific revolution seems to connote progress, while the other kind is seen as subversive or worse. But ultimately, of course, technology and science are simply manifestations of social behavior. The directions they take, the ends they serve, and the breakthroughs they achieve are shaped by the human purposes which guide them. If the social revolutions the world has thus far seen have proved to be neither breakthroughs nor adequately designed to meet the requirements of a humanistic, planetary social system, then these failures are failures of previous intellectual efforts, human will, and the shape of human values. They are not failures in the concept of breakthrough or the concept of revolution.

And so, implicit in the apparently simple initial question: "what is the environment and what is the system?" is another question which is not statable in the same simple terms. It begins with an evaluation of the human condition, or the condition of an individual, a group, or a class of persons according to some more or less explicit criteria. I, or we, experience life as "good," "indifferent," or "intolerable," and so on. And it is this evaluation that leads to the questions: "Should something be changed?" "Can something be changed?" "And if so, how?"

The Engineers of the Future

If engineers are to participate in the change process—if they are in fact going to be engineers according to the primitive definition of the term, if they are to be revolutionaries in the most productive and humanistic sense of the term—then they must learn to understand what definition of social system is implicit in any work with which they are engaged. They must learn to understand the more or less explicit perspectives from which their clients view the world, and make a decision about whether they wish to share this view of things and, if necessary, try to introduce more satisfactory perspectives.

For many firms and for many individuals, it may well be that traditional social system perspectives will require drastic modifications. It may well be that engineers, like many architects and other professional or technical workers who call themselves "advocate planners," will select their site of employment and their social system perspectives in terms of congenial value orientations and the existence of relevant *ad hoc* interest groups.

Technology is not an independent variable which proceeds like an impersonal force to shape the conditions of social life. Simple technological determinism is simply monumental "bunk." People decide what will happen to their environments, their living spaces, and their lives. And increasingly these decisions are shaped implicitly and explicitly through technological and scientific means.

And as engineers begin to understand more deeply the implications of their roles as designers, as well as occupants of social systems, they will begin to participate more fully in the decisions which will shape the conditions of life for themselves and everyone else on this planet.

BIBLIOGRAPHY

Bauer, R. A. (Ed.), *Social Indicators,* The M.I.T. Press, 1966.

Bell, Daniel (Ed.), *Toward the Year 2000: Work in Progress,* Beacon Press, 1969.

Boguslaw, Robert, *The New Utopians,* Prentice-Hall, 1965.

Boguslaw, Robert, "The Design Perspective in Sociology," in Wendell Bell and James A. Mau, *The Sociology of the Future,* Russell Sage Foundation, (1971), 240–258.

Boguslaw, Robert, and Robert H. Davis, "Social Process Modeling: A Comparison of a Live and Computerized Simulation," *Behavioral Science,* **14,** No. 3, (May 1969), 197–203.

Boguslaw, Robert, "Social Action and Social Change," in Erwin O. Smigel, *Handbook on the Study of Social Problems,* Rand McNally, (1971), 421–434.

Chase, Stuart, *The Proper Study of Mankind,* Harper and Row, 1956.

Fromm, Erich, *Marx's Concept of Man,* Frederick Unger, 1961.

Gross, Bertram M., *The State of the Nation: Social-System Accounting,* Tavistock, 1966.

Henry, Jules, *Culture Against Man,* Random House, Vintage Edition, 1965.

Kahn, Herman, and Anthony J. Wiener, *The Year 2000,* Macmillan, 1967.

Kolakowski, Leszek, *Toward a Marxist Humanism,* Grove Press, 1968.

Marcuse, Herbert, *An Essay on Liberation,* Beacon Press, 1969.

"The Port Huron Statement," in Paul Jacobs and Saul Landau, *The New Radicals: A Report with Documents,* Vintage Books, 1966.

Platt, John R., *The Step to Man,* Wiley, 1966.

Schaff, Adam, *A Philosophy of Man,* Delta Books, 1968.

Schaff, Adam, *Marxism and the Human Individual,* McGraw-Hill, 1970.

Sheldon, Eleanor Bernert, and Wilbert E. Moore, *Indicators of Social Change,* Russell Sage Foundation, 1968.

Simon, Herbert A., *Models of Man,* Wiley, 1957.

11

A Critique of the Systems Approach To Social Organizations

C. WEST CHURCHMAN

Professor of Business Administration
School of Business Administration
University of California, Berkeley

C. West Churchman is a Professor of Business Administration and an Associate Director of the Space Science Laboratory at the University of California, Berkeley. He is the Vice President of Research and Education for The Institute of Management Sciences and an editor for *Management Science* and *Philosophy of Science*. He is a co-author of one of the first introductory texts in operations research, *Introduction to Operations Research*, and has also authored *Theory of Experimental Inference, Methods of Inquiry, Prediction and Optimal Decision, The Design of Inquiring Systems, Challenge to Reason*, and *The Systems Approach*. *Challenge to Reason* was selected as one of the best management books of 1968 by the Academy of Management. *The Systems Approach* received the McKinsey Book Award for books on management.

This final chapter is a critique of the systems approach. Since what I am writing is largely philosophical in nature, I should explain exactly what philosophy is, once and for all. Philosophy is that activity which seeks difficulties in a serious enterprise. The philosopher is very unhappy if someone else sees how to solve these difficulties, because the philosopher really does not like solutions at all. My enterprise here is not to sell the systems approach, but to point out some of its fundamental difficulties. I should say that, although philosophers seek difficulties, they only seek them in those things they dearly love. Keep it in mind that this chapter will have no ending. It will certainly not arrive at a solution to the problems it poses, but on the other hand, it is given with a tone of a great deal of love and enthusiasm for this thing we are calling the systems approach.

This chapter also has another flavor to it. Some of the other chapters deal with large systems, such as computer systems and the systems the astronauts use. My emphasis is more on the social system or what we sometimes call the social organization.

Background

I am going to back up a little to give the setting and some of the history of the kinds of developments I want to discuss. Actually, if I were to look at the history in long range, I would certainly have to go back to the Greek city-states. Plato's *Republic*[1] is a good example of the systems approach to a government. Plato tries to identify the components and see how they are interrelated, and even to suggest that there are appropriate measures of performance associated with a society or organization.

As we come down through history we repeatedly see attempts to try to apply the systems approach, within the science and intellectual capabilities of the times, to social systems of various kinds. One of the very outstanding books on the systems approach is Jeremy Bentham's *An Introduction to the Principles of Morals and Legislation,*[2] which sets down not only the program for the systems approach, but the fundamental difficulties that I allude to later on.

But today it is not fashionable for the academic to cite things too far back in history—it makes him look as though he has not been keeping up with the recent journals. So I will just give you a very brief thumbnail sketch of organization theory from the 1930s to the present time. In the 1930s the interests of the social sciences in organization theory might be described in terms of the intuitive-normative. People were trying to take their experience with organizations and, together with some studies, develop what they called "principles of organization," of which perhaps the most famous is "the span of control." They went as far as to say that a manager should have no more than seven people reporting to him. If you are worrying about plus or minus epsilon, epsilon was about 1. That shows how refined this effort became.

In the 1940s there was a general disillusionment with the whole effort to develop the so-called principles of organization, and under the influence of positivism and other things, the social scientists decided to devote their efforts in large part to what one might call the descrip-

[1] Plato, *Republic,* translated into English by B. Jowett, Oxford, Clarendon Press, 1888.

[2] Jeremy Bentham, *An Introduction to the Principles of Morals and Legislation,* The Clarendon Press, 1907.

tions of organizations, without any commitment as to the value of the organization and its goals, or even to the way its business was run. These descriptions were based on interviews, questionnaires, and the like, which provided a great deal of valuable information about certain critical cases in the organization's life, but which really did not take the analysis of the organization and its structure very far at all. And indeed, a lot of that literature, as you read back over it, turns out to be pretty dull stuff simply because a lot of the things that bothered organizations in the 1940s do not bother us today.

Systems Analysis

The 1950s saw the advent of what we now call "systems analysis," "operations research," "management science," and a number of other names. The naming, I might say, causes difficulties for people who are not familiar with this development; the multitude of names was caused by a dislike of such names as "operations research" or even "systems analysis," and so other names were suggested. Unless you wish to be involved in the battle of names, you can assume that all of them refer to the same kind of effort, but with different emphases, depending on the speaker.

The Overall View

Systems analysis of the sort I am going to describe took off roughly around 1950, based on many prior attempts to perform the same task. Let us examine the logical steps of systems analysis of the 1950 model. You can imagine yourself as a member of a group which goes into an organization—it might be a business firm, a church, or a government agency—and you wish to try to help by using systems analysis or operations research. What you do, if you live in that time, is try to get an overview of the whole organization. We really liked this notion; we thought we were different from most people in life, because we looked at the whole system. We did not know exactly what we meant when we said that, and I point out later why we did not. We said, we are not interested in solving some specific problem the manager thinks he has. We are going to look at the whole organization, with his assistance, and after we understand the whole organization as a system, we will proceed to the specific problems. Thus your first objective is to get an overall picture of the organization, its problem areas, and where your analyses may be of some assistance.

A Feasible Problem

The next step is to find a "feasible" problem to work on. "Feasible" has to be in quotes because again we were not exactly sure what we meant by the label. But what it appeared to mean was that we felt we had certain capabilities of a mathematical or, more generally, of an "analytical" nature, and these capabilities could be applied to certain types of problems, but not to all problems. Analysis could not be applied, for example, to problems where the seriousness of the problem depended on personalities or politics, because we did not feel we had the capability of analyzing such problems, which were therefore "infeasible." But we could analyze problems such as the allocation of resources.

Thus, in the 1950s we saw many applications of operations research and systems analysis to problems like production control, because production is essentially an allocation problem. That is, the task is to figure out how much money and resources to put into the production of various kinds of products.

The Construction of a Model

So the 1950s saw the development of many production scheduling and inventory control models. It is important to note that these models incorporate a conflict of values within the organization. One objective of production control is to produce the items when they are needed. On the other hand, if you do that too successfully, you find you are building up inventories beyond your capabilities. Somehow or other you have to learn to balance those two conflicting interests on the part of the organization.

The 1950s also saw the extensive development of mathematical programming and waiting line problems, as well as game theory and simulation, which have received a great deal of publicity, but, in fact, were rarely applied to help solve real problems. I should comment on the structure of mathematical programming, because it will help in understanding the "analysis" of OR (operations research). The structure of a mathematical program is fairly simple. We have a measure z, which we call "the measure of performance" of a system. We assume that this measure is a function of a set of variables, $x_1, x_2, \ldots, x_i, \ldots, x_n$. These variables are supposed to represent levels of activity, within the organization, of the various divisions or departments, or whatever. The variables are under the control of some manager, which is a critical point as we will see later on. The manager can, so to speak, turn these knobs to whatever level he wants, including zero; when he wipes out department i he turns x_i down to zero.

We say that the aim of the management is to maximize this value z. In the simplest cases, z is a linear function of the x_i, and we call the analysis "linear programming." The important aspect of the model is that the capability of the manager with respect to controlling the variables is limited. We normally express this fact by saying that some function of the variables is bounded by an upper and a lower bound.

Just to impress you by size, one of the latest models of linear programming has 2 million variables, and some 35,000 constraint equations. Whatever you say about it, it is certainly the biggest piece of mathematics that has ever been done. You may wonder how one can possibly solve the problem, because if you have ever tried to find the maximum of a function subject to constraints, you will have noted that it is quite a laborious task even when you have just a few variables. The cleverness of the model building consists in developing techniques that enable one to save a lot of time, especially computer time, in grinding out a solution. The algorithms for solving both linear and nonlinear mathematical programs are quite impressive from the point of view of the complexity of the problem versus the time required to reach a solution.

But I am less interested in the beauties of the mathematics, than in the essential logic of the analysis which I now call to your attention. The logic says that it is possible to generate, at least for a section of the organization, not only the goals of the organization in terms of "profit maximization" or "satisfying public welfare," but to quantify these goals in terms of one unifying measure. This is to be done subject to certain constraints which are very closely related to the measure of performance. In fact, there is a method, which the logic supplies, that enables one to estimate what it costs to set a constraint at a certain level. In other words, the constraints are not necessarily set by some arbitrary management decisions. We as analysts can at least say what the value, or the negative value (cost), of setting a constraint in a certain manner actually is. So if the administration of Caltech says that in a given Division there shall be no more than x number of students and professors, then, in principle at least, using the logic of mathematical programming, it is possible to estimate the cost of such a constraint in terms of the basic underlying value z of the total system.

So far I have identified three stages of OR: the overall view, the identification of a feasible problem, and the construction of a model to try to handle that problem with the logic I have just outlined.

Data Gathering

The fourth step is to gather the data. The data in the case of a linear programming model essentially consists of estimating the coefficients. These coefficients tell us what an activity contributes to the score of the system (z).

I have found from my experiences with the sciences (and I have worked in the physical, as well as the social sciences) that data collection is a very curious activity from a philosophical point of view, in the sense that comparatively little is said about it, although it occupies the great majority of the research effort. This is especially true in OR. The students who graduate from OR programs believe that if they have the model in hand, certainly any decent organization will have the data in hand. That is tit for tat, is it not? So, for example, we have an inventory model. We require, as data, the cost of holding the inventory. Or, we have a waiting-line model, and we require the cost of waiting in line. The students expect to be able to find in the books of the company, or somewhere, what these costs actually are. Much to their amazement, nobody knows. But they may get some clues as to where to go and ask somebody; and, if they can communicate, they may get some answers.

For example, in my experience in the cost of holding inventory in the private sector, we have been given figures ranging from 3 to 36 percent per year. Both of those answers were based on what the answerer thought was a perfectly reasonable set of assumptions on his part. One answer was given by a company where the manager said, in effect, that if we hold money in inventory, then we cannot use it to invest in the acquisition of other companies. Now they would not acquire another company, unless that company could be shown to have a 36 percent return on their investment before taxes. Hence the cost of holding inventory is 36 percent of the value of the inventory. The other company was a holding company, a mother company so to speak; it loaned money to its divisions at 3 percent interest on the book. Hence the managers assumed that the cost of holding inventory was only 3 percent, because any time they wanted extra cash they could always go to "Mama" and she would give it to them at 3 percent interest per annum.

I am being a little bit facetious, of course, but the point is that an underlying difficulty has crept into the system. The difficulty is that you cannot acquire data about costs without understanding other aspects of the total system, namely, in this particular case, the financial policies with respect to liquid assets. In other words, even if we identify a feasible problem to be solved by inventory theory, we cannot solve that problem without going into the finance department and figuring out their policies with respect to the allocation of liquid assets.

This is an extremely important point throughout all of the studies that have been made in operations research. In order to acquire the basic information for the models, one has to make certain very strong assumptions about the characteristics of the larger system in which the particular problem being studied is embedded.

Another example of the same sort is the demand. Obviously, if we are going to build an inventory system for the production department, we have to understand what the pattern of demand is. Here the innocent student is apt to collect past data on demand, if it is available (and much to his surprise it usually is not). Using past records, he builds histograms and, because he understands statistics, he makes probability estimates of demand of various amounts. However, his entire strategy may be wrong, because the part of the system which is generating demand may be operating incorrectly. And, indeed, it often is. One can change the demand pattern in the private sector by changing advertising and pricing. So you cannot gather data on demand without assuming that the larger system is operating properly.

It is true of science in general that data collection always requires some assumptions as to the characteristics of the larger system. You cannot run an experiment on a laboratory table without making certain assumptions about what does and does not affect your results. And, of course, the assumptions may be wrong. The same thing is true in systems analysis; only the problem is magnified in our case, because of the extreme difficulty of trying to hold onto anything long enough to recognize it as a datum.

The Solution is Calculated

Well, by hook or crook, that is, by making assumptions, we gather the data, and with the data plugged into the model, we calculate a solution. The solution turns out to be something of the following kind: this is how resources ought to be allocated; this is how a particular production system ought to be scheduled; this is how a particular kind of transportation system should be designed.

Implementation

Beyond this point, which is the calculation of the solution, there is obviously another phase to come; in systems analysis we call it "implementation." Something has to happen to try to get the solution across. Interestingly enough, so enthusiastic were we in the 1950s, that we believed if we came out with a rational solution based on our model and on data which we thought was reasonably accurate, every manager would hug us, thinking it was really great that we had found the solution to his problem. And, if the managers did not implement our solution, then they were just stupid. So it is either a choice of accepting our solution or being accused of being stupid. We gradually began to realize that perhaps the implementation problem—how you go from conceptualization of the model to the actualization of the solution—is *the* most important problem of all; that what has come before is minor compared to the problem of how you change an organization in the light of analysis. And I would say that the 1960s have witnessed a sort of reversal of interest on the part of the OR practitioner; not on the part of the academic OR man, who threw up his hands in horror at the implementation problem, but the practitioner who had to worry about whether any of his work could be implemented.

To illustrate my point, in the 1950s I had a group of my graduate students write to authors of articles in *Operations Research,* which is one of two major journals in this country that prints articles in operations research. These authors had described studies that they had conducted. The students wrote to them and said, "We've read your very fascinating account of your study and we really think it is superb in every way; we're just curious to know what happened after you finished it." Every article ended with the calculation of a solution, but not one peep as to whether anyone had done anything about it. Only one of the respondents of the dozen or so replying was able to say that the solution had been implemented, or knew anything about the subsequent history of the study.

But, as I say, the practitioner of operations research developed in the 1960s a healthy respect for the implementation problem. They also developed a healthy respect in some cases for the managers that had failed to implement the OR solutions, because there was often some aspect of the situation that the OR practitioner had not realized was there.

Finally, if one is successful in implementation, or maybe if one is not, one recycles back through the system again, searching for another feasible problem, another model, another data collection, another calculation of the solution, and so forth. Needless to say, the steps I have described are not necessarily chronological, since, for example, the problem is usually reformulated many times before a "solution" is forthcoming. Thus most operations research groups act like doctors in the medical profession: when there are problems where they can help, they try to help solve them.

A Critique

The remainder of this chapter is a critique of this common procedure of systems analysis which I have just outlined. In making this critique I should say that I do it with a great deal of humility, because these steps are ones most of us have tried to teach in our operations research classes. They are not a bad view of how the practice goes on, but I think we are realizing in the 1970s that there are some good reasons to make us pause and ask whether we should not be searching for a better way to structure the activities of the systems approach.

The Measure of Performance

The people who have most critized operations research have usually attacked the concept of a unifying measure of performance of the system. They argue that it is not possible to find one goal of an organization. They may say that it is possible in the private sector the measure of performance is the net profit of the company, but in the nonmilitary public sector (such as libraries and hospitals), there is no such measure. But I think that this criticism of our effort to try to find a unifying measure of performance of the system missed the mark. We were not saying we were sure that we could always find this measure; for example, for the life of me I cannot even get a glimmer of what it is for a library. Many of my students have tried hard, but we have had to conclude we really do not know what a university library is trying to maximize. Some not so nice wags have rudely suggested it is the number of books retained on the shelves!

So we have to conclude that at the present time we do not understand enough about organizations like libraries, hospitals, and universities to be able to see how to generate a unifying measure of their performance. But this conclusion is characteristic of all science. There are many things that a given science is not able to measure, even in an approximate sense, although it knows about the existence of these things. Thus it is unfair to criticize us for trying to find a unifying measure of performance of the system on the grounds that it is extremely difficult to do it. Because we are latter-day rationalists, we argue that if organizations are to be rational in what they do, then when the managers make a decision they do in fact balance out the various conflicting values. Some of the very informative and exciting studies that have been conducted in operations research have been taken from past managerial decisions, and have shown, for example, what the managers were assuming the value of a life to be. A study during World War II on air pilots inferred that the value of a life, as estimated in decisions of safety by the Department of Defense, was approximately $100,000. That is in 1940 dollars; it would be much higher now if we used present-day dollars. We are often told that we cannot calculate the cost of a patient's waiting for medical service. The answer is that the calculation *is* done implicitly when you decide on a specific allocation of doctor's time; if the cost of waiting is implicitly assumed, then why not make it explicit?

Of course, there are other reasons why we cannot find a unifying measure, but I would rather continue by pointing out what I think at the present time are some of the things we have not done well in the systems approach. These sources of difficulty guarantee that we still have an exciting adventure ahead of us.

Who is the Client?

One of the sources of difficulty, I think, is the evidence we used for our data. I have already alluded to this when I said we went to see the various managers and tried to get estimates of things like the cost of holding inventory. We also used the managers very heavily in finding out which of the problems of the organization were the ones we should have been working on. In my language, we tended very strongly to assume that the real client of systems analysis is the administrator-manager. We were trying to ascertain from managers the information as to how the system should work, although we realized, of course, that in many systems the real clients are not the managers. In a hospital system, the *real* clients are surely the patients, or potential patients. In an educational system, the *real* clients are the people who have the need to be educated. And so on. We tended to take the managers as surrogates for the real clients.

I know that many people think that it is pretty obvious a systems analyst works for the guy who is going to pay him, and the guy who is going to pay him is the manager. What actually has happened in many cases is that this concentration of the systems analyst on the administrator-manager as his guide in seeking feasible solutions and in estimating the various other kinds of information has created what might be called an administrative screen between the systems analyst and the real client. Further, this screen often distorts the values of the real client.

For example, I am now involved in a study dealing with the utilization of information about earth resources from satellite, aircraft, and ground observations. Almost everyone in the study assumes that the "users" of earth resource information are people in some government agency. But government agencies are only intermediaries that are supposed to be serving the more ultimate clients. If these intermediaries are not serving the ultimate clients, then you see what happens to the systems analysis. We solve the wrong problem in a very precise and rigorous way.

In one of my seminars the students are asked to study a social system, and, using their best judgment, to identify (1) the decision maker, (2) the right client, and (3) the measure of performance as the decision maker recognizes it; and then to see whether in their judgment the system itself is serving the right client. They examine a wide variety of systems: a counterinsurgency group in the military, a hospital, a library, a department of a university, a private company, a club, a church, and so on. Last fall, 23 out of the 24 graduate students reported to the seminar that in their opinion the system was not serving the right client. They were not saying the managers were evil, but rather that the system had somehow drifted away from the service of the people who were rightfully supposed to be served by the system.

This is one of the challenges, then, that we face in systems analysis in the 1970s: to take much more seriously, than we have hitherto, the nature of the right client of the system and how we should go about trying to identify the client.

Two Approaches to Human Values

In all systems analysis, there is a moral overtone which also can be seen in planning and other activities where the intellectual becomes involved in trying to improve society. The moral overtone can be explained in terms of a book that Immanuel Kant wrote in the 1780s.[3] Kant argued that there are two approaches to human values. One is the prudential approach: its aim is to maximize what he called happiness, but what we would call utility, or the z-score. Essentially, said Kant, this approach ignores the value of an individual as an individual. It has to take individuals as collections of people with certain properties. It addresses itself to collective properties, and is incapable of considering people as unique individuals. Kant's was a very pro-

[3] Immanuel Kant, *Fundamental Principles of the Metaphysics of Morals,* translated by T. K. Abbott, London, Longmans Green, 1898.

phetic statement, because, if you look at the linear programming model, you will see that, if the x_i's referred not to individuals and manpower but to machines, the system would be described in just the same manner: the model for a machine would be the same as a model for a social system.

The other aspect of values, said Kant, deals with the individual. The prescription which Kant developed was that every individual ought to be treated as an end in himself and never as a means only. One aspect or difficulty of the systems approach and systems analysis is how to resolve this basic social conflict between the individual as an end-in-himself and the total social good. Systems analysts are do-gooders. They are trying to help improve social systems. But in all their studies they make recommendations which lead to treating individuals as means, rather than as ends. Hence in some sense they fail in that aspect of the total value system which deals with the moral dimension.

I can illustrate the last point by saying that in the 1950s, when we were trying to appraise the developments in operations research, we often said, "We could have saved such-and-such a company $3 million." "How could they fail to accept our solution?" Why not? Well, one reason is that the savings of the millions of dollars would have often meant firing hundreds of people. The solution meant treating people as means and not as ends in themselves.

The Pervasiveness of Systems

Closely related to this fundamental difficulty, and therefore an exciting prospect of systems science in the future, is the "political life" of the system. I have a theory from my experience with various kinds of systems, especially in the public sector, that there is developing today a pervasiveness of systems. Let me illustrate by discussing the educational system.

A traditional approach to the educational system is to say that it is divided into three components: administration, faculty, and students. Where these functions are fairly well delineated in the traditional university and college, the administrator has the function of trying to obtain funds, and of making the basic decisions as to their allocations in buildings, teaching programs, and the like; the faculty have the job of deciding what should be taught within the program; and the students have the job of taking what is taught and absorbing it, so to speak.

Now at many universities there has been a serious question about this traditional approach. In what I call the pervasiveness of the system, everyone thinks that he ought to be in the whole act—that education is a pervasive process, that it goes from womb to tomb—that everybody ought to be involved not only in deciding what is to be taught and to whom it should be taught, but also involved in the administration of the university. I see the same kind of thing gradually developing in the health system, where we are beginning to realize that perhaps some of the major problems of health are not disease problems. Health is a pervasive problem, especially if you live in a highly polarized community.

Another way to view pervasiveness is to point out that we are finding it more and more difficult to study adequately one sector of the social system at a time, in the manner I earlier characterized as the methodology of the 1950s. We are finding it very difficult, if not impossible, to study the problems of education independently from the problem of health, because education and health are nonseparable aspects of the social system. To try to set the system's boundaries in the "feasible" fashion may lead us to the study of precisely the wrong problem and the wrong answer, because we have failed to study the interrelationship of a particular problem with its ramifications in other domains of society.

Consider, for example, what the United Nations calls the "protein crisis." In the developing nations of the world, lack of protein in the diet is a very serious problem, most especially for the pregnant woman. In the United Nations studies of the protein problem, the emphasis has been on how to produce and distribute more protein. None of the studies seem to recognize that protein is an educational problem, because the importance of protein is not obvious to most people. The educational problem is just as much a part of the "protein crisis" as is its production.

Sometimes systems analysts believe they have adequately handled the pervasiveness of problems by studying "second order or *n*th order effects" of various kinds of decisions, much as a pebble thrown in a still pool creates *n*th order waves. But this is not what I am talking about at all. The *n*th order effects are important, to be sure, but I am arguing that we cannot determine whether to throw the pebble without making strong assumptions about the whole pool. We cannot say that such-and-such a decision is a "first approximation" without assuming the nature of the whole series.

A Systems Philosophy for the Future

Therefore I believe that systems analysis of the future will have to develop a different kind of philosophy from the one I described earlier. In the earlier philosophy one goes from solution to solution, improving the system here, and then there, and then over here, and so on. What I think we are going to have to do is adopt what might be called a "learning philosophy," and to drop the concept of "solutions."

What now happens is we come up with a solution and then we try to implement it, and if we do—fine; if we do not, we can go on to look for another problem. We are really not applying the systems approach to ourselves, because we are very opportunistic. The ambition would be that we would set forth, as best we can, a kind of systems approach to how we would proceed. We would not be dictated to by the emergency, or the most feasible problem, and so on. We would see that we had a whole program that we would work on. I do not mean that we would be better than anyone else in trying to lay out the total system we are engaged in, but we would use that as a basis of our own choices of what we work on, and then modify our strategies as we do with any systems approach.

We need a model of the systems approach if we are going to be honest. Essentially what must change is the notion that we arrive at solutions and then hunt for new problems. It is not that opportunistic. Of course what also must change, as I have already said, is our notion of the measure of performance and where we are going to get it from. So the strategy of the research work itself may change as we see that we cannot rely on a given sector of the organization to supply the information.

All this amounts to saying that we are going to have to apply the systems approach to systems analysis itself. It is incredible that this reflective concept has been so late in occurring. Operations research has very rarely been applied to operations research. Consequently, many operations research groups have failed in what they tried to do because they never looked at themselves from the point of view of their own principles. If they had, they might have begun to question some of the bases of what they were doing.

Now it has been suggested by the pragmatists that society learns how to improve itself only when it is bold enough to "experiment," that is, to accept the fact that it does not know the answers and is willing to test some suggestions that come out of analysis and other methods of planning to see whether they "work out." But this is a very simplistic view of learning, because it assumes that we have the right perspective for telling whether something "works out." If a policy produces wealth for many and starvation for a few, the majority will say it "worked out." Thus we must recognize that when we try to see whether a given plan has worked in practice, we have to do this within a certain framework, which may be wrong. We have to be capable of holding different representations of social reality in hand at the same time.

What is needed, then, is a dialectical approach to systems improvement. Otherwise, we are apt to find ourselves bound to one representation of social reality, and this will prevent our being able to learn from the kinds of activities we are engaged in. I would say the keynote concept of the systems approach for the future will be the dialectical learning process. Having said that, I should also admit I do not know exactly what that process means; the essence of education is the ability to recognize what you do not know, but need to know.

BIBLIOGRAPHY

Churchman, C. West, *Challenge to Reason,* McGraw-Hill, 1968.

Churchman, C. West, *The Design of Inquiring Systems,* Basic Books, 1971.

Churchman, C. West, *Prediction and Optimal Decision,* Prentice-Hall, 1961.

Churchman, C. West, *The Systems Approach,* Delacorte Press, 1968.

Churchman, C. West, *Theory of Experimental Inference,* Macmillan, 1948.

Churchman, C. West, and R. L. Ackoff, *Methods of Inquiry,* Educational Publishers, 1950.

Churchman, C. West, R. L. Ackoff, and E. L. Arnoff, *Introduction to Operations Research,* Wiley, 1957.

Churchman, C. West, and Philburn Ratoosh, *Measurement: Definitions and Theories,* Wiley, 1959.

12

Bibliography

Abadie, J. (Ed.) *Nonlinear Programming,* Wiley, 1967.

Ackoff, Russell L., *Scientific Method: Optimizing Applied Research Decisions,* Wiley, 1962.

Ackoff, Russell L., and Patrick Rivett, *A Manager's Guide to Operations Research,* Wiley, 1963.

Ackoff, R. L., and M. W. Sasieni, *Fundamentals of Operations Research,* Wiley, 1970.

Alexander, J. E., and J. M. Bailey, *Systems Engineering Mathematics,* Prentice-Hall, 1962.

Allison, David, *The R&D Game,* The M.I.T. Press, 1969.

The Analysis and Evaluation of Public Expenditures: The PPB System, Vols. I, II, and III, U.S. Government Printing Office, 1969.

Anstadter, Bertram L., *Reliability Mathematics,* McGraw-Hill, 1971.

Aoki, Masanao, *Optimization of Stochastic Systems,* Academic Press, 1967.

Aoki, M., *Introduction to Optimization Techniques,* Macmillan, 1971.

Arrow, K. J., *Social Choice and Individual Values,* Wiley, 1951.

Asimow, Morris, *Introduction to Design,* Prentice-Hall, 1962.

Athens, Michael, and Peter L. Falb, *Optimal Control,* McGraw-Hill, 1966.

Ayres, Robert U., *Technological Forecasting and Long-Range Planning,* McGraw-Hill, 1969.

Barlow, R. E., and F. Proschan, *Mathematical Theory of Reliability,* Wiley, 1965.

Barron, F., W. C. Dement, W. Edwards, J. Olds, M. Olds, and T. M. Newcomb, *New Directions in Psychology,* Vol. 2, Holt, Rinehart, and Winston, 1965.

Barton, Richard F., *A Primer on Simulation and Gaming,* Prentice-Hall, 1970.

Barucha-Reid, A. T., *Markov Processes and Their Applications,* McGraw-Hill, 1960.

Bauer, Raymond A. (Ed.), *Social Indicators,* The M.I.T. Press, 1966.

Bauer, Raymond A., and Kenneth J. Gergan, *The Study of Policy Formation,* Macmillan, 1968.

Bauer, Raymond A., et al., *NASA Planning and Decision Making,* Vols. I and II, Harvard Graduate School of Business Administration, NGR 22-007-163, 1969.

Baumgartner, John S., *Project Management,* Irwin, 1963.

Baumol, William J., *Economic Theory and Operation Analysis,* Prentice-Hall, 1965.

Bazovsky, Igor, *Reliability Theory and Practice,* Prentice-Hall, 1961.

Bedini, Silvio, Wernher von Braun, and Fred L. Whipple, *Moon: Man's Greatest Adventure,* Abrams, 1970.

Bell, Daniel (Ed.), *Toward the Year 2000: Work in Progress,* Beacon Press, 1969.

Bellman, Richard, *Dynamic Programming,* Princeton University Press, 1957.

Bellman, Richard, and Robert Kalaba, *Dynamic Programming and Modern Control Theory,* Academic Press, 1965.

Bellman, Richard E., and Stuart E. Dreyfus, *Applied Dynamic Programming,* Princeton University Press, 1962.

Berrien, F. K., *General and Social Systems,* Rutgers University Press, 1968.

Bertalanffy, Ludwig von, *General System Theory,* Braziller, 1968.

Bertalanffy, Ludwig von, and Anatol Rapoport (Ed.), *General Systems: Yearbook of the Society for the Advancement of General Systems,* Vol. I, Braun-Brunfield, 1956.

A Bibliography of Selected Rand Publications: Systems Analysis, The Rand Corporation, SB-1022, 1967.

Black, Guy, *Application of Systems Analysis to Government Operations,* Praeger, 1968.

Blanchard, B., and E. Lowery, *Maintainability Principles and Practices,* McGraw-Hill, 1969.

Blaquiere, Austin, *Nonlinear System Analysis,* Academic Press, 1966.

Blesser, William B., *A Systems Approach to Biomedicine,* McGraw-Hill, 1969.

Blumenthal, Sherman C., *Management Information Systems,* Prentice-Hall, 1969.

Boguslaw, Robert, *The New Utopians,* Prentice-Hall, 1965.

Borch, Karl H., *The Economics of Uncertainty,* Princeton University Press, 1968.

Boyd, Dean W., *A Methodology for Analyzing Decision Problems Involving Complex Preference Assessments,* Decision Analysis Group, Stanford Research Institute, May 1970.

Braybrooke, David, and Charles E. Lindblom, *A Strategy of Decision,* Free Press of Glencoe, 1963.

Bright, James R. (Ed.), *Technological Forecasting for Industry and Government,* Prentice-Hall, 1968.

Brockett, Roger W., *Finite Dimensional Linear Systems,* Wiley, 1970.

Bross, Irwin D. J., *Design for Decision,* Macmillan, 1953.

Brown, B. M., *The Mathematical Theory of Linear Systems,* Chapman and Hall, 1965.

Brown, R., E. Galanter, E. H. Hess, and G. Mandler, *New Directions in Psychology,* Vol. 1, Holt, Rinehart, and Winston, 1962.

Bryson, Arthur E., and Yu-Chi Ho, *Applied Optimal Control,* Blaisdell, 1969.

Buckley, Walter (Ed.), *Modern System Research for the Behavioral Scientist,* Aldine, 1968.

Buckley, Walter, *Sociology and Modern Systems Theory,* Prentice-Hall, 1967.

Canon, Michael D., Clifton D. Cullum, and Elijah Polak, *Theory of Optimal Control and Mathematical Programming,* McGraw-Hill, 1970.

Cetron, Marvin J., *Technological Forecasting: A Practical Approach,* Gordon and Breach, 1969.

Cetron, Marvin J., Raymond Isenson, et al., *Technical Resource Management,* The M.I.T. Press, 1969.

Chacko, George K., *Applied Statistics in Decision-Making,* Elsevier, 1971.

Chartrand, Robert L., *Systems Technology Applied to Social and Community Problems,* Spartan Books, 1971.

Chase, Stuart, *The Proper Study of Mankind,* Harper and Row, 1956.

Chestnut, Harold, *Systems Engineering Methods,* Wiley, 1967.

Chestnut, Harold, *Systems Engineering Tools,* Wiley, 1965.

Chorafas, Dimitris N., *Systems and Simulation,* Academic Press, 1965.

Churchman, C. West, *Challenge to Reason,* McGraw-Hill, 1968.

Churchman, C. West, *The Design of Inquiring Systems,* Basic Books, 1971.

Churchman, C. West, *Prediction and Optimal Decision,* Prentice-Hall, 1961.

Churchman, C. West, *The Systems Approach,* Delacorte Press, 1968.

Churchman, C. West, *Theory of Experimental Inference,* Macmillan, 1948.

Churchman, C. West, and R. L. Ackoff, *Methods of Inquiry,* Educational Publishers, St. Louis, 1950.

Churchman, C. West, R. L. Ackoff, and E. L. Arnoff, *Introduction to Operations Research,* Wiley, 1957.

Churchman, C. West, and Philburn Ratoosh, *Measurement: Definitions and Theories,* Wiley, 1959.

Cleland, David I., *Systems Analysis and Project Management,* McGraw-Hill, 1968.

Cleland, David I., and William R. King, *Systems, Organization, Analysis, Management: A Book of Readings,* McGraw-Hill, 1969.

Cohen, Jacob W., *The Single Server Queue,* North-Holland, 1969.

Conway, R. W., W. L. Maxwell, and L. W. Miller, *Theory of Scheduling,* Addison-Wesley, 1967.

Cooper, Leon, and David Steinberg, *Introduction to Methods of Optimization,* Saunders, 1970.

Corliss, W. R., *Space Probes and Planetary Exploration,* Van Nostrand, 1965.

Cox, D. R., *Renewal Theory,* Wiley, 1962.

Cox, D. R., and H. D. Miller, *The Theory of Stochastic Processes,* Wiley, 1965.

Dantzig, George B., *Linear Programming and Extensions,* Princeton University Press, 1963.

Dean, Burton V. (Ed.), *Operations Research in Research and Development,* Wiley, 1963.

DeGreene, Kenyon B. (Ed.), *Systems Psychology,* McGraw-Hill, 1970.

Derusso, Paul M., Rob J. Roy, and Charles M. Close, *State Variables for Engineers,* Wiley, 1965.

Deutsch, Ralph, *System Analysis Techniques,* Prentice-Hall, 1969.

Dewey, John, *How We Think,* D. C. Heath, 1910.

Dixon, J. R., *Design Engineering: Inventiveness, Analysis, and Decision Making,* McGraw-Hill, 1966.

Dole, S. H., et al., *Establishment of A Long-Range Planning Capability,* The Rand Corporation, RM-6151-NASA, NAS 2-5459, September 1969.

Dole, S. H., et al., *Methodologies for Analyzing the Comparative Effectiveness and Costs of Alternate Space Plans,* RM-5656-NASA (Vols. 1 and 2), The Rand Corporation, 1968.

Dorfman, R., P. A. Samuelson, and R. M. Solow, *Linear Programming and Economic Analysis,* McGraw-Hill, 1958.

Drake, A. W., *Fundamentals of Applied Probability Theory,* McGraw-Hill, 1967.

Drummond, M., *System Evaluation and Measurement Techniques,* Prentice-Hall, 1971.

Duckworth, W. E., *A Guide to Operational Research,* Methuen, 1962.

Easton, David, *A Systems Analysis of Political Life,* Wiley, 1965.

Eckman, D. P. (Ed.), *Systems: Research and Design,* Wiley, 1961.

Eddison, R. T., K. Pennycuick, and B. M. P. Rivett, *Operational Research in Management,* Wiley, 1962.

Edwards, Ward, and Amos Tsversky (Ed.), *Decision Making,* Penguin Books, 1967.

Edwards, Ward, "Decision Making is Too Simple for Executives to Waste Time on It," *Innovation,* No. 5, (1969), 34–41.

Edwards, Ward, et al., "Probabilistic Information Processing Systems: Design and Evaluation," IEEE Transactions on *Systems Science and Cybernetics,* Vol. SSC-4, No. 3, September 1968.

Elements of Design Review for Space Systems, NASA SP-6502, National Aeronautics and Space Administration, 1967.

Elgerd, Olle I., *Control Systems Theory,* McGraw-Hill, 1967.

Ellis, David O., and Fred J. Ludwig, *Systems Philosophy,* Prentice-Hall, 1962.

Emery, F. E. (Ed.), *Systems Thinking,* Penguin Books, 1969.

Emery, J. C., *Organizational Planning and Control Systems,* Macmillan, 1969.

English, J. M., *Economics for the Practicing Engineer,* Barnes and Noble, 1971.

English, J. Morley (Ed.), *Cost-Effectiveness,* Wiley, 1968.

English, J. Morley, *Economics of Engineering and Social Systems,* Wiley, 1972.

Enke, Stephen (Ed.), *Defense Management,* Prentice-Hall, 1967.

Enthoven, Alain C., and K. Wayne Smith, *How Much Is Enough*?, Harper and Row, 1971.

Epstein, Richard A., *The Theory of Gambling and Statistical Logic,* Academic Press, 1967.

Farquharson, Robin, *Theory of Voting,* Yale University Press, 1969.

Feinberg, Gerald, *The Prometheus Project,* Doubleday, 1968.

Feller, William, *An Introduction to Probability Theory and Its Applications,* Wiley, Vol. I, 1950; Vol. II, 1966.

Fellner, William, *Probability and Profit: A Study of Economic Behavior Along Bayesian Lines,* Irwin, 1965.

Finkelstein, R. P., et al., *Methods for the Selection of NASA Programs and Projects,* Stanford Research Institute, TR 7988-1, NAS 2-5455, August 1969.

Finley, Dorothy L., et al., *Human Performance Prediction in Man-Machine Systems,* NASA CR-1614 (Prepared by Bunker-Ramo Corp.), National Aeronautics and Space Administration, August 1970.

Fishburn, Peter C., *Decision and Value Theory,* Wiley, 1964.

Fishburn, Peter C., *Utility Theory for Decision Making,* Wiley, 1970.

Fisher, Gene H., *Cost Considerations in Systems Analysis,* American Elsevier, 1970.

Flagle, C. D., W. H. Huggins, and R. H. Roy, *Operations Research and Systems Engineering,* The Johns Hopkins Press, 1960.

Fogel, Lawrence, *Biotechnology: Concepts and Applications,* Prentice-Hall, 1963.

Ford, L. R., and D. R. Fulkerson, *Flows in Networks,* Princeton University Press, 1962.

Formby, John, *An Introduction to the Mathematical Formulation of Self-Organizing Systems,* Van Nostrand, 1965.

Forrester, J. W., *Industrial Dynamics,* The M.I.T. Press, 1961.

Forrester, J. W., *Principles of Systems,* Wright-Allen Press, 1968.

Forrester, J. W., *Urban Dynamics,* The M.I.T. Press, 1969.

Forrester, J. W., *World Dynamics,* Wright-Allen Press, 1971.

Fox, J. (Ed.), *System Theory,* Polytechnic Press, 1965.

Frank, H., and Ivan T. Frisch, *Communications, Transmission and Transportation Networks,* Addison-Wesley, 1971.

Freeman, Herbert, *Discrete-Time Systems,* Wiley, 1965.

Gagne, R. M., *Psychological Principles in System Development,* Holt, Rinehart, and Winston, 1962.

Gale, David, *The Theory of Linear Economic Models,* McGraw-Hill, 1960.

Garvin, Walter W., *Introduction to Linear Programming,* McGraw-Hill, 1960.

Gass, Saul I., *An Illustrated Guide to Linear Programming,* McGraw-Hill, 1970.

Gass, Saul I., *Linear Programming,* McGraw-Hill, 1969.

Goldman, A. S., and T. B. Slattery, *Maintainability—A Major Element of System Effectiveness,* Wiley, 1964.

Goldman, Thomas A. (Ed.), *Cost-Effectiveness Analysis: New Approaches in Decision Making,* Praeger, 1967.

Goode, Harry H., and Robert E. Machol, *System Engineering,* McGraw-Hill, 1957.

Gordon, Geoffrey, *System Simulation,* Prentice-Hall, 1969.

Gosling, W., *The Design of Engineering Systems,* Wiley, 1962.

Gray, W., N. D. Rizzo, and F. J. Duhl (Ed.), *General Systems Theory and Psychiatry,* Little, Brown, Boston, 1967.

Greenberg, Harold, *Integer Programming,* Academic Press, 1971.

Greenwood, Frank, *Managing the Systems Analysis Function,* American Management Association, 1968.

Gross, Bertram M., *The State of the Nation: Social-System Accounting,* Tavistock, 1966.

A Guide to System Engineering, Army Technical Manual TM 38–760, Department of Defense, April 1969.

Hadley, G., *Introduction to Probability and Statistical Decision Theory,* Holden-Day, 1967.

Hadley, G., *Linear Programming,* Addison-Wesley, 1962.

Hadley, G., *Nonlinear and Dynamic Programming,* Addison-Wesley, 1964.

Hahn, G. J., and S. S. Shapiro, *Statistical Models for Systems Engineering,* Wiley, 1967.

Hajek, Victor G., *Project Engineering,* McGraw-Hill, 1965.

Hall, Arthur D., *A Methodology for Systems Engineering,* Van Nostrand, 1962.

Hansen, B. J., *Practical PERT Including Critical Path Method,* America House, 1964.

Hare, Jr., Van Court, *Systems Analysis: A Diagnostic Approach,* Harcourt, Brace and World, 1967.

Hartman, W., H. Matthes, and A. Proeme, *Management Information Systems Handbook,* McGraw-Hill, 1968.

Haugen, E. G., *Probabilistic Approaches to Design,* Wiley, 1968.

Helmer, Olaf, *Analysis of the Future: The Delphi Method,* The Rand Corporation, P-3558, March 1967.

Helmer, Olaf, *Social Technology,* Basic Books, 1966.

Hillier, F. S., and G. J. Lieberman, *Introduction to Operations Research,* Holden-Day, 1967.

Hinrichs, H. H., and G. M. Taylor, *Program Budgeting and Benefit-Cost Analysis,* Goodyear Publishing Co., 1969.

Hitch, Charles J., *Decision-Making for Defense,* University of California Press, 1965.

Hitch, Charles J., and Roland N. McKean, *The Economics of Defense in the Nuclear Age,* Harvard University Press, 1960.

Hogg, Robert V., and Allen T. Craig, *Introduction to Mathematical Statistics,* Macmillan, 1965.

Hovey, Harold A., *The Planning-Programming-Budgeting Approach to Government Decision Making,* Praeger, 1968.

Howard, Ronald A., *Dynamic Probabilistic Systems,* Vol. 1: *Markov Models,* Vol. 2: *Semi-Markov and Decision Processes,* Wiley, 1971.

Howard, Ronald A., *Dynamic Programming and Markov Processes,* The M.I.T. Press, 1960.

Howard, Ronald A. (Ed.), "Special Issue on Decision Analysis," IEEE Transactions on *Systems Science and Cybernetics,* Vol. SSC-4, No. 3, September 1968.

Hu, T. C., *Integer Programming and Network Flows,* Addison-Wesley, 1971.

Hyvarinen, L. P., *Information Theory for Systems Engineers,* Springer-Verlag, 1970.

Iberall, Arthur, *Toward a General Science of Viable Systems,* McGraw-Hill, 1972.

Introduction to the Derivation of Mission Requirements Profiles for System Elements, NASA SP-6503, National Aeronautics and Space Administration, 1967.

An Introduction to the Evaluation of Reliability Programs, NASA SP-6501, Office of Technology Utilization, National Aeronautics and Space Administration, 1967.

Ireson, W. Grant (Ed.), *Reliability Handbook,* McGraw-Hill, 1966.

Jaiswal, N. K., *Priority Queues,* Academic Press, 1968.

Jantsch, Erich, *Technological Forecasting in Perspective,* Organization for Economic Cooperation and Development, Paris, 1967.

Johnson, R. A., F. E. Kast, and J. E. Rosenzweig, *The Theory and Management of Systems,* McGraw-Hill, 1967.

Kahn, Herman and Anthony J. Wiener, *The Year 2000,* Macmillan, 1967.

Kahn, Herman, and Irwin Mann, *Ten Common Pitfalls,* The Rand Corporation, RM-1937, July 1957.

Kalman, R. E., P. L. Falb, and M. A. Arbib, *Topics in Mathematical System Theory,* McGraw-Hill, 1969.

Kalmus, H., *Regulation and Control in Living Systems,* Wiley, 1966.

Karlin, Samuel, *A First Course in Stochastic Processes,* Academic Press, 1966.

Kast, Fremont E., and James E. Rosenzweig, *Science, Technology, and Management,* McGraw-Hill, 1963.

Kaufmann, A., *Methods and Models of Operations Research,* Prentice-Hall, 1963.

Kelleher, Grace J. (Ed.), *The Challenge to Systems Analysis: Public Policy and Social Change,* Wiley, 1970.

Kepner, Charles H., and Benjamin B. Tregoe, *The Rational Manager,* McGraw-Hill, 1965.

Kirk, D. E., *Optimal Control Theory, An Introduction,* Prentice-Hall, 1970.

Klir, George J., *An Approach to General Systems Theory,* Van Nostrand Reinhold, 1969.

Klir, George J. (Ed.), *Trends in General Systems Theory,* Wiley, 1972.

Kupperman, Robert H., and Harvey A. Smith, *Mathematical Foundations of Systems Analysis,* Addison-Wesley, 1969.

Lasdon, Leon S., *Optimization Theory for Large Systems,* Macmillan, 1970.

Lass, Harry, and Peter Gottlieb, *Probability and Statistics,* Addison-Wesley, 1971.

Laszlo, Ervin, *Introduction to Systems Philosophy,* Gordon and Breach, 1971.

Lawrence, J. R. (Ed.), *Operations Research and the Social Sciences,* Tavistock Publications, 1966.

Lay, Beirne, Jr., *Earthbound Astronauts: The Builders of Apollo Saturn,* Prentice-Hall, 1971.

Lazzaro, Victor, *Systems and Procedures,* Prentice-Hall, 1968.

Lee, Wayne, *Decision Theory and Human Behavior,* Wiley, 1971.

Leitmann, G. (Ed.), *Optimization Techniques with Applications to Aerospace Systems,* Academic Press, 1962.

Lifson, M. W., *Decision and Risk Analysis,* Barnes and Noble, 1971.

Lindblom, Charles E., *The Intelligence of Democracy: Decision Making Through Mutual Adjustment,* Free Press of Glencoe, 1965.

Lindley, D. V., *Introduction to Probability and Statistics from a Bayesian Viewpoint,* Cambridge University Press, 1965.

Lloyd, David K., and Myron Lipow, *Reliability: Management, Methods, and Mathematics,* Prentice-Hall, 1962.

Logsdon, John M., *The Decision to Go to the Moon,* The M.I.T. Press, 1970.

Lorens, C. S., *Flowgraphs: For the Modeling and Analysis of Linear Systems,* McGraw-Hill, 1964.

Lott, Richard W., *Basic Systems Analysis,* Harper and Row, 1971.

Luce, R. D., *Individual Choice Behavior,* Wiley, 1959.

Luce, R. Duncan, and Howard Raiffa, *Games and Decisions,* Wiley, 1957.

Luenberger, David G., *Optimization by Vector Space Methods,* Wiley, 1969.

Lyden, F. J., and E. G. Miller (Ed.), *Planning Programming Budgeting: A Systems Approach to Management,* Markham, 1967.

McKean, Roland N., *Efficiency in Government Through Systems Analysis,* Wiley, 1958.

McKean, Roland N., *Public Spending,* McGraw-Hill, 1968.

McMillan, Claude, Jr., *Mathematical Programming: An Introduction to the Design and Application of Optimal Decision Machines,* Wiley, 1970.

McMillan, Claude, *Systems Analysis: A Computer Approach to Decision Models,* Irwin, 1968.

MacCrimmon, K. R., *Decision Making Among Multiple-Attribute Alternatives: A Survey and Consolidated Approach,* RM-4823-ARPA, The Rand Corporation, December 1968.

Machol, Robert E. (Ed.), *System Engineering Handbook,* McGraw-Hill, 1965.

Machol, Robert E. (Ed.), *Information and Decision Processes,* McGraw-Hill, 1960.

Machol, Robert E., and Paul Gray (Ed.), *Recent Developments in Information and Decision Processes,* Macmillan, 1962.

Makower, M. S., and E. Williamson, *Operational Research,* The English Universities Press, 1967.

The Management Process for Development of Army Systems, Army AR 11-25, Department of Defense, April 10, 1968.

Management Study of NASA Acquisition Process, National Aeronautics and Space Administration, June 1971,

Mandler, G., P. Mussen, N. Kogan, and M. A. Wallach, *New Directions in Psychology,* Vol. 3, Holt, Rinehart, and Winston, 1967.

Mangasarian, O., *Nonlinear Programming,* McGraw-Hill, 1968.

March, James G., and Herbert A. Simon, *Organizations,* Wiley, 1958.

Martin, J. J., *Bayesian Decision Problems and Markov Chains,* Wiley, 1967.

Matthews, Don Q., *The Design of the Management Information System,* Auerbach, 1971.

Mesarovic, M. D. (Ed.), *System Theory and Biology,* Springer-Verlag, 1968.

Mesarovic, M. D., *Views on General Systems Theory,* Wiley, 1964.

Mesarovic, M. D., D. Macko, and Y. Takahara, *Theory of Hierarchical, Multilevel, Systems,* Academic Press, 1970.

Meyer, P. L., *Introductory Probability and Statistical Applications,* Addison-Wesley, 1965.

Miller, David W., and Martin K. Starr, *Executive Decisions and Operations Research,* Prentice-Hall, 1969.

Miller, Robert W., *Schedule, Cost, and Profit Control with PERT,* McGraw-Hill, 1963.

Milsum, John H., *Biological Control Systems Analysis,* McGraw-Hill, 1966.

Moder, J. J., *Project Management with CPM and PERT,* Van Nostrand Reinhold, 1970.

Morris, William T., *Management Science: A Bayesian Introduction,* Prentice-Hall, 1968.

Morse, Philip M., *Library Effectiveness: A Systems Approach,* The M.I.T. Press, 1968.

Morse, Philip M., *Queues, Inventories and Maintenance,* Wiley, 1958.

Morse, Philip M., and George E. Kimball, *Methods of Operations Research,* Wiley, 1951.

Morse, Philip M., and Laura W. Bacon (Ed.), *Operations Research for Public Systems,* The M.I.T. Press, 1967.

Morton, J. A., *Organizing for Innovation: A Systems Approach to Technical Management,* McGraw-Hill, 1971.

Morton, M. S., *Management Decision Systems: Computer-Based Support for Decision Making,* Harvard Business School, 1971.

NASA PERT and Companion Cost System Handbook, National Aeronautics and Space Administration, October 30, 1962.

National Goals Research Staff, *Toward Balanced Growth: Quantity with Quality,* U.S. Government Printing Office, 1970.

Naylor, T. H., et al., *Computer Simulation Techniques,* Wiley, 1966.

Nemhauser, George L., *Introduction to Dynamic Programming,* Wiley, 1966.

Novick, David, *Program Budgeting: Program Analysis and the Federal Budget,* Harvard University Press, 1967.

O'Brien, James J., *Management Information Systems,* Van Nostrand Reinhold, 1970.

Optner, Stanford L., *Systems Analysis for Business Management,* Prentice-Hall, 1960.

Optner, Stanford L., *Systems Analysis for Business and Industrial Problem Solving,* Prentice-Hall, 1965.

Papoulis, Athanasios, *Probability, Random Variables, and Stochastic Processes,* McGraw-Hill., 1965.

Parzen, Emanuel, *Modern Probability Theory and Its Applications,* Wiley, 1960.

Patten, Bernard C. (Ed.), *Systems Analysis and Simulation in Ecology,* Academic Press, 1971.

Peck, Merton J., and Frederick M. Scherer, *The Weapons Acquistion Process: An Economic Analysis,* Harvard University Press, 1962.

Pelz, Donald C., and Frank M. Andrews, *Scientists in Organizations,* Wiley, 1966.

Peterson, E. L., *Statistical Analysis and Optimization of Systems,* Wiley, 1961.

Platt, John R., *The Step to Man,* Wiley, 1966.

Polak, E., *Computational Methods in Optimization,* Academic Press, 1971.

Polya, G., *How to Solve It,* Doubleday, 1957.

Popper, Karl R., *The Logic of Scientific Discovery,* Harper and Row, 1965.

Porter, William A., *Modern Foundations of Systems Engineering,* Macmillan, 1966.

Prabhu, Narahari U., *Queues and Inventories: A Study of their Basic Stochastic Processes,* Wiley, 1965.

Pratt, John W., Howard Raiffa, and Robert O. Schlaifer, *Introduction to Statistical Decision Theory,* McGraw-Hill, 1965.

Progress in Operations Research, Vol. I: R. L. Ackoff (Ed.), 1961; Vol. II: D. B. Hertz and R. T. Eddison (Eds.), 1964; Vol. III: J. S. Aronofsky (Ed.), 1969; Wiley.

Quade, E. S. (Ed.), *Analysis for Military Decisions,* Rand McNally, 1965.

Quade, E. S., and W. I. Boucher, *Systems Analysis and Policy Planning,* American Elsevier, 1968.

Raiffa, Howard, *Decision Analysis,* Addision-Wesley, 1968.

Raiffa, Howard, *Preferences for Multi-Attributed Alternatives,* RM-5868-DOT/RC, The Rand Corporation, April 1969.

Raiffa, Howard, and Robert Schlaifer, *Applied Statistical Decision Theory,* The M.I.T. Press, 1961.

Ramo, Simon, *Century of Mismatch,* McKay, 1970.

Ramo, Simon, *Cure for Chaos,* McKay, 1969.

Ramo, Simon, "New Dimenstions of Systems Engineering," *Science and Technology in the World of the Future,* A. B. Bronwell (Ed.), Wiley, 1970, pp. 127–146.

Rau, John G., *Optimization and Probability in Systems Engineering,* Van Nostrand Reinhold, 1970.

Riley, Vera, and Saul I. Gass, *Linear Programming and Associated Techniques,* (Bibliography), The Johns Hopkins Press, 1958.

Riordan, J., *Stochastic Service Systems,* Wiley, 1962.

Rivlin, Alice, *Systematic Thinking for Social Action,* The Brookings Institution, 1971.

Roberts, Edward B., *The Dynamics of Research and Development,* Harper and Row, 1964.

Rothenberg, Jerome, *The Measurement of Social Welfare,* Prentice-Hall, 1961.

Rubio, J. E., *The Theory of Linear Systems,* Academic Press, 1971.

Rudwick, Bernard H., *Systems Analysis for Effective Planning,* Wiley, 1969.

Saaty, Thomas L., *Elements of Queueing Theory,* McGraw-Hill, 1961.

Saaty, Thomas L., *Mathematical Methods of Operations Research,* McGraw-Hill, 1959.

Saaty, Thomas L., *Mathematical Models of Arms Control and Disarmament,* Wiley, 1968.

Sackman, H., *Man-Computer Problem Solving,* Auerbach Publishers, 1970.

Sackman, Harold, *Computers, System Science, and Evolving Society,* Wiley, 1967.

Sage, Andrew P., *Optimum Systems Control,* Prentice-Hall, 1968.

Sage, Andrew P., and James L. Melsa, *System Identification,* Academic Press, 1971.

Samaras, T. T., and F. L. Czerwinski, *Fundamentals of Configuration Management,* Wiley, 1971.

Sandler, G. H., *System Reliability and Engineering,* Prentice-Hall, 1963.

Sasieni, M., A. Yaspan, and L. Friedman, *Operations Research—Methods and Problems,* Wiley, 1959.

Savage, L. J., *The Foundations of Statistics,* Wiley, 1954.

Sayles, Leonard R., and Margaret K. Chandler, *Managing Large Systems,* Harper and Row, 1971.

Schelling, Thomas C., *The Strategy of Conflict,* Harvard University Press, 1960.

Scherer, F. M., *The Weapons Acquisition Process: Economic Incentives,* Harvard Business School, 1964.

Schlaifer, Robert, *Analysis of Decisions Under Uncertainty,* McGraw-Hill, 1969.

Schlaifer, Robert, *Introduction to Statistics for Business Decisions,* McGraw-Hill, 1961.

Schultz, Charles L., *Setting National Priorities: The 1971 Budget,* The Brookings Institution, 1970.

Schwarz, Ralph J., and Bernard Friedland, *Linear Systems,* McGraw-Hill, 1965.

Seiler, John A., *Systems Analysis in Organizational Behavior,* Dorsey Press, 1967.

Seiler, Karl, *Introduction to Systems Cost Effectiveness,* Wiley, 1969.

Sheldon, Alan, Frank Baker, and Curtis P. McLaughlin (Eds.), *Systems and Medical Care,* The M.I.T. Press, 1970.

Shelly, Maynard W., and Glenn L. Bryan (Ed.), *Human Judgements and Optimality,* Wiley, 1965.

Shinners, Stanley M., *Techniques of System Engineering,* McGraw-Hill, 1967.

Shooman, M. L., *Probabilistic Reliability: An Engineering Approach,* McGraw-Hill, 1968.

Siegel, Sidney, *Choice, Strategy and Utility,* McGraw-Hill, 1964.

Simon, Herbert, *The New Science of Management Decision,* Harper and Row, 1960.

Simon, Herbert A., *Models of Man,* Wiley, 1957.

Simonnard, Michel, *Linear Programming,* Prentice-Hall, 1966.

Singh, Jagjit, *Great Ideas of Operations Research,* Dover, 1968.

Spivey, W. Allen, and Robert M. Thrall, *Linear Optimization,* Holt, Rinehart, and Winston, 1970.

Stark, Robert M., and Robert L. Nicholls, *Mathematical Foundations for Design: Civil Engineering Systems,* McGraw-Hill, 1971.

System Engineering, U.S. Army Management Engineering Training Agency, August 1970.

System Engineering Management for Defense Material Items, DOD MIL STD 499, Department of Defense.

System Engineering Management Procedures, Air Force Systems Command Manual (AFSCM 375-5), United States Air Force, February 1964.

System/Project Management, DOD Directive 5010.14, Department of Defense, May 4, 1965.

The Systems Approach to Management (An Annotated Bibliography), NASA SP-7501, National Aeronautics and Space Administration, 1969.

Systems Engineering in Space Exploration, Jet Propulsion Laboratory, 1965.

Takacs, L., *Introduction to the Theory of Queues,* Oxford University Press, 1962.

Takacs, Lajos, *Combinatorial Methods in the Theory of Stochastic Processes,* Wiley, 1967.

Thiel, H., John C. G. Boot, and Teun Kloek, *Operations Research and Quantitative Economics,* McGraw-Hill, 1968.

Thierauf, Robert J., and Richard A. Grosse, *Decision Making through Operations Research,* Wiley, 1970.

Thrall, R. M., C. H. Coombs, and R. L. Davis (Eds.), *Decision Processes,* Wiley, 1954.

Tocher, K. D., *The Art of Simulation,* Van Nostrand, 1963.

Tribus, Myron, *Rational Descriptions, Decisions, and Designs,* Pergamon, 1969.

Truxal, John G., *Introductory Systems Engineering,* McGraw-Hill, 1972.

Tucker, Samuel A. (Ed.), *A Modern Design for Defense Decision: A MacNamara-Hitch-Enthoven Anthology,* Industrial College of the Armed Forces, 1966.

Vadja, S., *An Introduction to Linear Programming and the Theory of Games,* Methuen and Science Paperbacks, 1966.

Vadja, S., *The Theory of Games and Linear Programming,* Methuen, 1956.

Vazsonyi, Andrew, *Scientific Programming in Business and Industry,* Wiley, 1958.

von Alven, William H. (Ed.), *Reliability Engineering,* Prentice-Hall, 1964.

von Neumann, John, and Oskar Morgenstern, *Theory of Games and Economic Behavior,* Wiley, 1944.

Wagner, Harvey M., *Principles of Operations Research with Applications to Managerial Decisions,* Prentice-Hall, 1969.

Walton, Thomas F., *Technical Data Requirements for Systems Engineering and Support,* Prentice-Hall, 1965.

Weapon System Effectiveness Advisory Committee (WSEIAC), Final Report of Task Group II, AFSC-TR-65-2, United States Air Force, January 1965.

Weatherall, M., *Scientific Method,* Simon and Schuster, 1968.

Webb, James E., *Space Age Management,* McGraw-Hill, 1969.

"What Made Apollo A Success?", *Astronautics and Aeronautics,* **8,** No. 3 (March 1970), Special Section.

White, H. J., *Systems Analysis,* Saunders, 1969.

Wiener, Anthony J., et al., *A Workbook of Alternative Future Environments for NASA Mission Analysis,* Hudson Institute, HI-1271/2-IR, NAS 2-5431, December 31, 1969.

Wilde, Douglass J. and Charles Beightler, *Foundations of Optimization,* Prentice-Hall, 1967.

Wilford, John Noble, *We Reach the Moon,* Bantam Books, 1969.

Williams, J. D., *The Compleat Strategyst,* McGraw-Hill, 1954.

Wilson, Jr., E. Bright, *An Introduction to Scientific Research,* McGraw-Hill, 1952.

Wilson, I. G., and M. E. Wilson, *Information, Computers, and System Design,* Wiley, 1965.

Wilson, I. G., and M. E. Wilson, *Management, Innovation, and System Design,* Auerbach, 1971.

Wilson, Warren E., *Concepts of Engineering System Design,* McGraw-Hill, 1965.

Wismer, David A. (Ed.), *Optimization Methods for Large Scale Systems,* McGraw-Hill, 1972.

Wong, Eugene, *Stochastic Processes in Information and Dynamical Systems,* McGraw-Hill, 1972.

Wymore, A. Wayne, *A Mathematical Theory of Systems Engineering: The Elements,* Wiley, 1967.

Zadeh, L. A., and C. A. Desoer, *Linear System Theory: The State Space Approach,* McGraw-Hill, 1963.

Zadeh, L. A., and E. Polak, *System Theory,* McGraw-Hill, 1969.

Zahradnik, Raymond L., *Theory and Techniques of Optimization for Practicing Engineers,* Barnes and Noble, 1971.

Zangwill, Willard I., *Nonlinear Programming,* Prentice-Hall, 1969.

Zwicky, Fritz, *Discovery, Invention, Research,* Macmillan, 1966.

Zwicky, F., and A. G. Wilson (Eds.), *New Methods of Thought and Procedure,* Springer-Verlag, 1967.

Index

Aggregation rule, 88, 94
Apollo, 7, 10, 148, 151–164
  command module, 153
  flight design, 157
  fire, 162
  lunar module (LM), 153
  lunar surface environment, 156
  meteoroid hazard, 154
  radiation hazard, 153
  reliability, 161
  testing the Saturn V, 159

Bayes's Theorem, 88, 92, 95, 99, 107
Bayes, Thomas, 93
Beyesian, processor, 96
  technology, 102

California Institute of Technology (Caltech), 91, 93, 125, 163, 173, 195
Centralization, principle of, 39
Choice, 53, 57, 60, 61
Clairvoyant, 65
Corporal missile, 126
Cost/effectiveness, 38

Decision, definition, 53, 65
Decision analysis, 6, 51–85, 87–110
Decision analysis cycle, 68
Decision analysis examples, medical, 61, 74, 100
  nuclear reactor, 79
  police department, 105
  power system, 69
  weather, 74, 102
  world war, 98
Decision-making, 6, 39, 51–85, 87–110, 165
Decision theory, 55

Descriptive models, 40
Design, cycle, 36, 38
  interface, 137
  simplicity, 137
  for success, 136
  systems, 22, 36
Diagnosis, 88, 89, 93

Effectiveness, 5, 34, 38, 47, 115
Encoding of information, 58
Engineer, compleat systems, 48
  definition, 151
Engineering, definition, 178
  as a profession, 177
Environment, 34, 51, 188, 190
Escalation ladder, U.S. and Soviet Union, 187
Evaluation, 88, 102
Events of low probability, 39
Expected value, 38, 63, 96

Game theory, 48, 186

Human functions, 180
Human values, individual approach to, 202
  prudential approach to, 201

Implementation, 1, 37, 132, 135, 147, 197
Inference, 94
Information, 54, 89, 107, 138
  encoding of, 58
  processing, 99, 107, 109
  processors, 94
  value of, 65
Ingenuity, 53
Inputs, 34, 41, 180
Intangibles, 77

Interfaces, 34, 36, 134
Intuition, 53
Inventory models, 115, 118, 194, 196

Jet-assisted takeoff (JATO), 126
Jet Propulsion Laboratory (JPL), 7, 125–150
Judgment, 53, 87, 88, 105, 148

Launch vehicle, 129, 134, 138, 157
Likelihood ratio, 92, 96, 98
Lunar surface environment, 156

Management science, 193
Markov models, 115
Martin Luther King, Jr. General Hospital, 145
M.I.T. Operations Research Center, 118
Matrix organizational structure, 132
Medicare and Medicaid, 166
Mission plan, 138
Models, 40, 58, 111
  decision analysis, 58
  descriptive, 40
  dynamic programming, 45
  hospital admitting system, 146
  inventory, 116, 118, 194, 196
  linear and nonlinear, 42
  linear programming, 44, 117, 195
  Markov, 115
  mutual bargaining, 172
  network flow, 45
  normative, 40
  optimization, 44
  scheduling, 194
  simulation, 40
  structural/empirical, 40
  symbolic, 40
  of systems approach, 204
Morgantown Project, 143, 149

National Aeronautics and Space Administration (NASA), 7, 37, 129, 139, 148 151
New York City-Rand Institute, 172
Normative models, 40

Objective function, 44, 117, 174
Operations research, 6, 111–124, 193
Operations research example, Army inventory, 118
  blood inventory, 119
  libraries, 116, 199
  petroleum company, 116
  ship unloading, 114
  traffic, 121
  World War II, 46, 114
Optimization, 2, 44, 149
Organization theory, 8, 192
Organization, principles of, 192
Outcomes, 53, 54, 88
Outputs, 34, 41, 71, 180

Perception, 53
Planning-Programming-Budgeting Systems (PPBS), 7, 165–175, 183
  analysis of issues for, 171
  component parts of, 170
  essential aspects of, 170
  obstacles to, 168
  origins of, 170
  presenting plans for, 171
  systematic analysis with, 172
Poisson distribution, 118
POGO, 159
Policy, 73, 109, 116, 122
Preferences, 54, 60, 78, 174
Probability, 58, 64, 77, 83, 88, 90–96, 107, 108
Probabilistic Information Processing System (PIP), 96, 107
Program Evaluation and Review Technique (PERT), 45, 142
Programming, dynamic models, 45, 117
  linear models, 44, 117
  mathematical models, 44, 117
Project, lunar and planetary, 129, 132, 139
  management, 127, 139, 148
  Morgantown, 143–145, 149
  organization, 148
Project Development Plan, 139

Queuing, models, 113

Redundancy, 137
Reliability, 35, 129, 139, 140, 181
Risk, 61, 63, 148

Saturn V., 159–161
Schedule, 139, 140
Science, 4, 5, 14, 185

method of, 111, 122
Sergeant missile, 128
Simulation, 15, 40, 117, 121, 174, 183
Social value function, 71
Social welfare, 4
Space program, future of, 163
Space shuttle, 163
Spacecraft, 129–142
State, system, 42
Strategy, thermonuclear war, 186
Strategy of conflict, 187
Suboptimization, 39, 46, 115
Subsystem, 35, 134, 135, 139, 142, 143, 188
Surveys, public opinion, 123
Systems, analysis, 9, 31, 36, 132, 145, 193
approach, 1, 8–11, 13–32, 132, 147, 191–205
automobile, 19
chronological phases of design, 36
city as, 23
civil, 4, 5, 14, 35, 143–147
characteristics of, 3, 34
compleat engineer, 48
component of, 36, 37
computer, 26
conceptual subdivision of, 36
crime reduction, 184
definition, 2, 33
design, 22, 36, 38
dialectical approach to, 205
economic, 29
educational, 5, 182, 202
engineering, 1, 2, 33–50, 125–150, 151
health care delivery, 145
hierarchy, 4
hospital, 24
library, 10, 116, 199
nation-state, 188
pervasiveness of, 202
Planning-Programming-Budgeting (PPBS), 7, 165–175, 183
principles of design, 38
Probabilistic Information Processing (PIP), 96
social, 4, 5, 7, 35, 177–190, 191–205
social versus nonsocial, 181
state, 42
team, 8, 16, 126
technical, 4, 5, 35
telephone, 17
transportation, 4, 21, 25, 35
urban health, 145

Technology, 4, 5, 185, 190
Testing, 138
Saturn V., 159
Trade-offs, 15, 25, 35, 37, 38, 62, 103, 132, 183

Uncertainty, 52, 58
Urban Health Systems Task, 145
Urban Mass Transit Administration (UMTA), 143
Utility, 88, 103

Value, 61, 88, 102, 181, 188, 189, 195, 200, 201, 202
expected, 38, 63, 96
of information, 65
of life, 64, 66, 78, 183, 200

Warfare, systems analysis of, 186
Welare, 4, 167, 184